AF759933

Applications of Gas Chromatography

Applications of Gas Chromatography

Editor

Adrianna Coty

scitus academics

Applications of Gas Chromatography
Edited by **Adrianna Coty**

ISBN: 978-1-68117-262-0
Library of Congress Control Number: 2016937430

© 2017 by
SCITUS Academics LLC,
www.scitusacademics.com
Box No. 4766, 616 Corporate Way,
Suite 2, Valley Cottage,
NY 10989

This book contains information obtained from highly regarded resources. Copyright for individual articles remains with the authors as indicated. All chapters are distributed under the terms of the Creative Commons Attribution License, which permits unrestricted use, distribution, and reproduction in any medium, provided the original author and source are credited.

Notice

Reasonable efforts have been made to publish reliable data and views articulated in the chapters are those of the individual contributors, and not necessarily those of the editors or publishers. Editors or publishers are not responsible for the accuracy of the information in the published chapters or consequences of their use. The publisher believes no responsibility for any damage or grievance to the persons or property arising out of the use of any materials, instructions, methods or thoughts in the book. The editors and the publisher have attempted to trace the copyright holders of all material reproduced in this publication and apologize to copyright holders if permission has not been obtained. If any copyright holder has not been acknowledged, please write to us so we may rectify.

Printed in United States of America on Acid Free Paper

Preface

Gas chromatography is a term used to describe the group of analytical separation techniques used to analyze volatile substances in the gas phase. In gas chromatography, the components of a sample are dissolved in a solvent and vaporized in order to separate the analytes by distributing the sample between two phases: a stationary phase and a mobile phase. The mobile phase is a chemically inert gas that serves to carry the molecules of the analyte through the heated column. Gas chromatography is one of the sole forms of chromatography that does not utilize the mobile phase for interacting with the analyte. The stationary phase is either a solid adsorbant, termed gas-solid chromatography (GSC), or a liquid on an inert support, termed gas-liquid chromatography (GLC). Helium remains the most commonly used carrier gas in about 90% of instruments although hydrogen is preferred for improved separations. This inert gas goes through a glass column packed with silica that is coated with a liquid. Materials that are less soluble in the liquid will increase the result faster than the material with greater solubility. The purpose of this book entitled Applications of Gas Chromatography is to provide a better understanding on its separation and measurement techniques and its application. Since chromatography techniques are separating and analyzing methods, this book will help other researchers and young scientists to choose a suitable chromatography technique. Furthermore, this book illustrates the newest challenges in this area. This valuable book aims to provide a connection between various chromatography techniques and different processes.

Table of Contents

CHAPTER 1

Applications of Solid-Phase Microextraction and Gas Chromatography/Mass Spectrometry (SPME-GC/MS) in the Study of Grape and Wine Volatile Compounds

Annarita Panighel and Riccardo Flamini

Consiglio per la Ricerca e la Sperimentazione in Agricoltura-Centro di Ricerca per la Viticoltura (CRA-VIT), Viale XXVIII aprile 26, Conegliano (TV) 31015, Italy

ABSTRACT

Volatile compounds are responsible for the wine "bouquet", which is perceived by sniffing the headspace of a glass, and of the aroma component (palate-aroma) of the overall flavor, which is perceived on drinking. Grape aroma compounds are transferred to the wine and undergo minimal alteration during fermentation (e.g., monoterpenes and methoxypyrazines); others are precursors of aroma compounds which form in winemaking and during wine aging (e.g., glycosidically-bound volatile compounds and C_{13}-norisoprenoids). Headspace solid phase microextraction (HS-SPME) is a fast and simple technique which was developed for analysis of volatile compounds. This review describes some

SPME methods coupled with gas chromatography/mass spectrometry (GC/MS) used to study the grape and wine volatiles.

KEYWORDS

grape; wine; aroma; volatile compounds; SPME; GC/MS

1. INTRODUCTION

Wine aroma is formed by more than 800 volatile compounds and is characteristic of each product [1,2]. Often these compounds are present in very low concentration and are characterized by very low sensory thresholds (between ng/L and μg/L). Usually, the wine aroma profiling needs a sample preparation for isolation and concentration of the volatiles before performing gas chromatographic analysis. Several sample preparation methods for the analysis of grapes and wine were proposed: distillation [3,4], liquid–liquid extraction (LLE) [5,6], solid phase extraction (SPE) [7,8], dynamic headspace extraction [9], and headspace–solid phase microextraction (HS-SPME) [10,11,12,13].

SPME was developed in the 1990s by Pawliszyn and co-workers [14] and every year a thousand papers describing different aspects of this approach, and applications in different fields (chemical analysis, bioanalysis, food science, environmental science, and recently, pharmaceutical and medical sciences), are published [15]. This sample extraction technique was demonstrated to be rapid, simple, and reproducible, with no solvent use, and is suitable for the extraction and concentration of a high number of volatile and semi-volatile compounds from aqueous solutions [16,17]. Moreover, SPME needs a small sample volume and the coupling with gas chromatography and mass spectrometry (GC/MS) provides high sensitivity. For these reasons it has been used to study the volatile profile of many fruit varieties, vegetables, and beverages, including grapes and wine [18,19,20].

This paper reviews the main SPME-GC/MS applications developed to study the volatile and aroma compounds of grapes and wine.

1.1. Grape and Wine Volatile Compounds

Since early 80's a great number of studies of grape and wine volatiles have been performed and the main compounds identified are listed in Table 1. Principal in grape are monoterpenes, C_{13}-norisoprenoids, benzene compounds, C_6aldehydes, and alcohols [21,22,23]. These compounds are present in berry skin and pulp in both free (volatile) and glycosidically-bound (non-volatile) form. Free volatile compounds directly contribute to grape and wine aroma while glycosides are flavorless compounds which can act as aroma precursors for enzymatic and acid hydrolysis occurring in winemaking and during wine storage [24].

In general, monoterpenes have floral and citrus notes; C_{13}-norisoprenoids such as β-damascenone and β-ionone are characterized by "fruity-flowery", "honey-like", "sweet" and "violet" notes, respectively [25].

Main grape and wine volatile benzenoids are aldehydes and alcohols, such as benzaldehyde, phenylacetaldehyde (hyacinth and rose-like odor [26]), vanillin, benzyl alcohol, and 2-phenylethanol [24], probably formed from L-phenylalanine via the shikimic pathway [27]. The structures of the main grape aroma compounds are shown in Figure 1,Figure 2 and Figure 3.

Twenty-four aldehydes (mostly alkyls) were identified in wine; they can form during winemaking or be present in the grapes [28]. For example, glyoxal, methylglyoxal, hydroxypropandial and glycolaldehyde form by microorganisms such as*Saccharomyces cerevisiae* or *Leuconostoc oenos*, or can form in grapes as a consequence of *Botrytis cinerea* grape attack [30,31]. In general, C_6 aldehydes (hexanal, (*E*)-2-hexenal, and (*Z*)-3-hexenal) and (*Z*)-2-nonenal are responsible for the green, herbaceous, and sometimes bitter aroma of wines [21]. They are mainly due to the enzymatic cleavage of oxidized linoleic and linolenic acid during grape crushing before fermentation [29].

Table 1. Principal aroma compounds identified in grapes and wine [20,23,24,28,29].

Terpenoids	**Benzenoids**	**Sulfur compounds**
linalool	zingerone	methyl mercaptan
nerol	zingerol	ethyl mercaptan
geraniol	acetophenone	dimethyl sulfide
citronellol	vanillin	diethyl sulfide
α-terpineol	methyl salicylate	dimethyl disulfide
cis/*trans* ocimenol	eugenol	diethyl disulfide
cis/*trans* linalooloxide (furanic form)	*cis*/*trans* isoeugenol	methyl thioacetate
cis/*trans* linalooloxide (pyranic form)	2-phenylethanol	ethyl thioacetate
hydroxycitronellol	benzyl alcohol	2-mercaptoethanol
8-hydroxydihydrolinalool	acetovanillone	2-(methylthio)-1-ethanol
7-hydroxygeraniol	benzaldehyde	3-(methylthio)-1-propanol
7-hydroxynerol	4-hydroxybenzaldehyde	4-(methylthio)-1-butanol
cis/*trans* 8-hydroxylinalool	2,4-dimethylbenzaldehyde	2-furanmethanethiol
diendiol I	phenylacetaldehyde	benzothiazole
endiol	syringaldehyde	thiazole
diendiol II	coniferaldehyde	5-(2-hydroxyethyl)-4-methylthiazole
neroloxide	sinapaldehyde	4-methyl-4-mercaptopentan-2-one
2-*exo*-hydroxy-1,8-cineol	propriosyringone	3-mercaptohexanol acetate
1,8-cineol	propriovanillone	*cis*/*trans* 2-methylthiophan-3-ol
cis/*trans* 1,8-terpine	syringol	2-methyltetrahydrothiophen-3-one
p-menthenediol I	coniferyl alcohol	*cis*/*trans* 2-methyltetrahydrothiophen-3-ol
(*E*)-geranic acid	vanillic alcohol	3-mercaptohexan-1-ol
(*E*)-2,6-dimethyl -6-hydroxyocta-2,7-dienoic acid	sinapic alcohol	3-mercaptohexyl acetate
(*E*)- and (*Z*)-sobrerol	*o*-cymene	4-mercapto-4-methylpentan-2-ol
cis/*trans* rose oxide	*p*-cymene	3-mercapto-3-methylbutan-1-ol
lilac alcohols	guaiacol	
triol	4-ethylguaiacol	
hotrienol	4-vinylguaiacol	
myrcenol	4-ethylphenol	
limonene	4-vinylphenol	
β-phellandrene	methyl anthranilate	
β-ocimene	2'-aminoacetophenone	
wine lactone		
Aliphatic alcohols	**Acids**	**Sesquiterpenes**
1-butanol	isobutyric acid	rotundone
2-nonanol	isovaleric acid	farnesol
3-methyl-1-butanol	acetic acid	germacrene D
2-methyl-1-butanol	butyric acid	γ-cadinene
isobutanol	hexanoic acid	α-ylangene
1-pentanol	octanoic acid	α-farnesene
1-hexanol	decanoic acid	β-farnesene
1-octanol	hexadecanoic acid	nerolidol
(*E*)-3-hexen-1-ol	octadecanoic acid	
(*Z*)-3-hexen-1-ol		
4-methyl-3-penten-1-ol		
(*E*)-2-hexen-1-ol		
1-octen-3-ol		
2-ethyl-1-hexanol		
furfuryl alcohol		
6-methyl-5-hepten-2-ol		

Table 1. *Cont.*

Carbonyl compounds	**Esters**	**Norisoprenoids**
acetaldehyde	ethyl 2-methylpropanoate	TDN (1,1,6-trimethyl-1,2-dihydronaphthalene)
isobutyraldehyde	ethyl 2-methylbutanoate	β-damascone
2-methylbutanal	ethyl 3-methylbutanoate	β-damascenone
isovaleraldehyde	ethyl 2-hydroxypropanoate	vomifoliol
1-octen-3-one	ethyl 3-hydroxybutanoate	dihydrovomifoliol
(*E*)-2-heptenal	ethyl 4-hydroxybutanoate	3-hydroxy-β-damascone
methional	diethyl succinate	3-oxo-α-ionol
(*E*)-2-octenal	diethyl malate	3-hydroxy-7,8-dihydro-β-ionol
hexanal	ethyl butanoate	α-ionol
(*E*)-2-hexenal	ethyl hexanoate	β-ionol
(*Z*)-3-hexenal	ethyl octanoate	α-ionone
(*Z*)-2-nonenal	ethyl decanoate	β-ionone
furfural	ethyl benzoate	actinidols
5-methylfurfural	isoamyl octanoate	vitispiranes
1H-pyrrole-2-carboxyaldehyde	ethyl furoate	Riesling acetal
geranial	ethyl dihydrocinnamate	hydroxy-megastigmen-2-one
neral	ethyl cinnamate	hydroxy-megastigmen-3-one
acetoin	methyl vanillate	4-oxo-isophorone
diacetyl	ethyl vanillate	β-isophorone
glyoxal	ethyl acetate	4-oxo-2,3-dehydro-β-ionol
methylglyoxal	isobutyl acetate	β-cyclocitral
glycolaldehyde	isoamyl acetate	
hydroxypropandial	ethyl 2-phenylacetate	
2,4-nonadienal	hexyl acetate	
2,6-nonadienal		

Lactones	**Nitrogen compounds**	
γ-butyrolactone	3-isobutyl-2-methoxypyrazine	
γ-hexalactone	3-sec-butyl-2-methoxypyrazine	
γ-nonalactone	3-isopropyl-2-methoxypyrazine	
γ-decalactone	3-ethyl-2-methoxypyrazine	
cis/*trans* oak lactone		
sotolon		

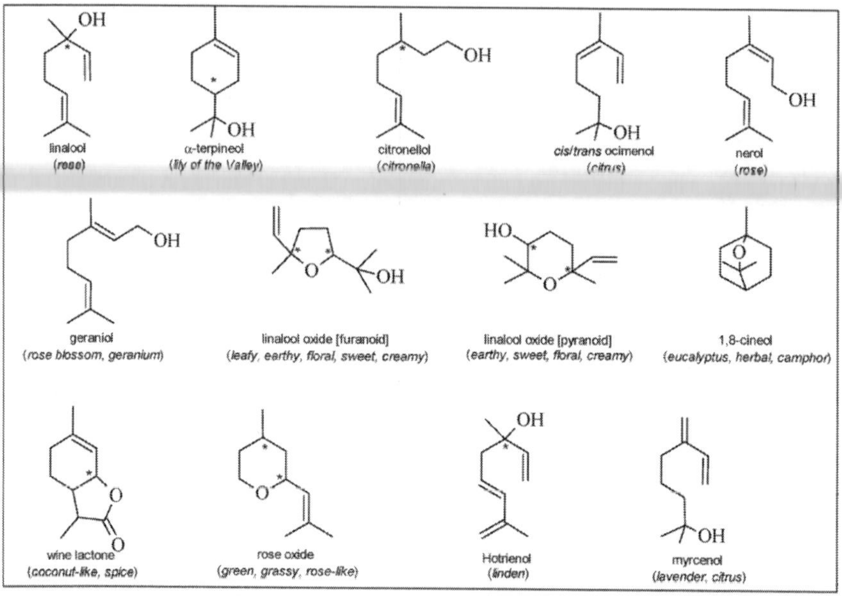

Figure 1. Principal monoterpenes identified in grapes. In brackets, the odor descriptor is reported [24].

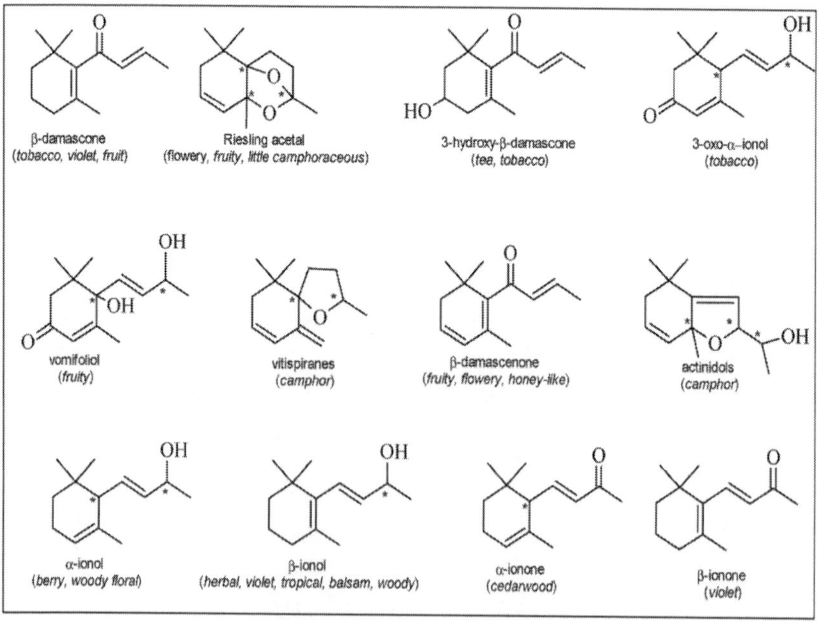

Figure 2. Principal C_{13}-norisoprenoids identified in grapes. In brackets, the odor descriptor is reported [24].

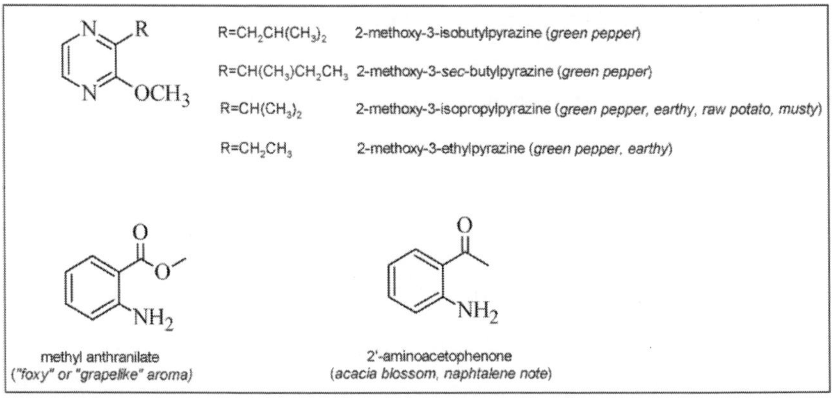

Figure 3. Structures of "foxy-smelling" compounds and alkyl methoxypyrazines in grapes. In brackets, the odor descriptor is reported [25].

Main wine volatiles are ethyl esters, acetates, and higher alcohols. Esters are produced by the yeasts during fermentation. Principal are ethyl esters characterized by fruity and floral notes, such as ethyl hexanoate, ethyl octanoate, ethyl decanoate, ethyl dodecanoate, isoamyl acetate, hexyl acetate, and 2-phenylethyl acetate [32,33]. Contents of hexanoic, octanoic, and decanoic acid in wine depend on the yeast strain, fermentation conditions and grape must composition [27]. Higher alcohols are formed by the yeast sugar metabolism (anabolic pathway) as well as *via* the catabolic or Ehrlich pathway of amino acids [27,34,35]. Rapp and Versini reported that a higher alcohols concentration below 300 mg/L is desirable for the aroma complexity of wine whereas a concentration exceeding 400 mg/L can have a detrimental effect [36].

Volatile phenols, such as 4-vinylphenol (aroma descriptors: spicy, pharmaceutical), 4-vinylguaiacol (smoke, phenolic), 4-ethylphenol (horse stable, medicinal), and 4-ethyl guaiacol (spice, phenolic), are produced during alcoholic fermentation by some microorganisms (*Brettanomyces* yeasts and bacteria) by decarboxylation of hydroxycinnamic acids present in must [32,33,37].

2. SPME-GC/MS METHODS

2.1. Analysis of PFBOA-Derivatives

In general, carbonyl compounds contribute to the wine aroma even though they are present in low levels. GC/MS analysis of the *O*-pentafluorobenzyl (PFB) derivatives formed by reaction with *O*-(2,3,4,5,6-pentafluorobenzyl)-hydroxylamine (PFBOA) by recording in singular-ion-monitoring (SIM) the mass spectrum base peak signal at m/z 181 (characteristic of PFB-oximes) is a selective and sensitive method. On the other hand, derivatization increases the complexity of analyte peaks in the chromatogram due to formation of two oximes for each carbonyl group (isomers *E* and *Z*, except for formaldehyde) [38]. By this method several studies of carbonyl compounds in hydro-alcoholic matrices, such as wine, model wine solutions, and spirits, were performed [30,31,39,40,41,42, 43,44,45,46,47].

Malolactic fermentation (MLF) is an important oenological process performed after the alcoholic fermentation for improving the organoleptic characteristics and the microbiological stability of wine [46]. The process is carried out by lactic acid bacteria: it can occur naturally or be induced by the inoculum of commercial bacteria strains. Conversion of L(−)-malic into L(+)-lactic acid decreases wine acidity [30]; usually, the inoculation of selected bacteria strains enables control over the process [48].

With MLF, together with the other fermentative compounds (esters, sulfur and nitrogen compounds, volatile phenols, and volatile fatty acids), also the carbonyl profile changes, increasing the aromatic complexity of wine [42,49].

HS-SPME and GC/MS analysis of PFBOA derivatives was performed to study the carbonyl compounds in Merlot wines after MLF [46]. In general, PFBOA derivatives are characterized by higher volatility, thermal stability, and affinity for the SPME fiber with respect to the corresponding carbonyl compounds [20]. SPME was performed by using a polydimethylsiloxane/divinylbenzene (PDMS/DVB) fiber; GC/MS analysis was performed by electron impact ionization (EI) and chemical ionization in both positive (PICI) and negative (NCI) mode [50]. NCI coupled to SPME provided lower detection limits for isobutyraldehyde, 2-methylbutanal, isovaleraldehyde, (*E*)-2-hexenal, 1-octen-3-one, (*E*)-2-heptenal, methional, (*E*)-2-octenal, phenylacetaldehyde, and (*E*)-2-nonenal [50]. PICI was used for determination of the principal carbonyl compounds in wine, such as acetaldehyde, diacetyl, and acetoin [46]. By using methane as reagent gas, abundant formation of acetaldehyde derivative $[M+H]^+$ ion at *m/z* 240, diacetyl mono-derivative $[M+H]^+$ ion at *m/z* 282, and *o*-chlorobenzaldehyde derivative $[M+H]^+$ ion at *m/z* 336 (the internal standard), was observed. For acetoin abundant formation of $[M+H-18]^+$ ion at *m/z* 266 was observed. These signals were used for the quantification of compounds. Experimental PICI conditions used are reported in Table 2 [46].

2.2. Aroma Compounds and Wine Aging

Aging in wooden barrels is a process used to stabilize the color and to improve limpidity and the sensorial characteristics of wines. During aging many compounds are transferred from the wood to the wine: polyphenols, lactones, coumarins, polysaccharides, hydrocarbons and fatty acids, terpenes, C_{13}-norisoprenoids, steroids, carotenoids, and furan compounds. Wood volatiles such as *cis-* and *trans*-β-methyl-γ-octalactones (oak lactones), syringaldehyde, vanillin, coniferaldehyde, sinapaldehyde and their alcohols, propiosyringone and propiovanillone, hydroxy-megastigmen-2-one and hydroxy-megastigmen-3-one [51], furfural, 5-methyl furfural, guaiacol, eugenol, 4-ethylphenol and 4-ethyl guaiacol [52,53], furfuryl alcohol, β-ionone, γ-noanalactone, and acetovanillone [54] confer the typical organoleptic characteristics of aged wines.

Table 2. Experimental conditions used for PFBOA derivatization and SPME-GC/MS analysis (ion trap and positive chemical ionization) of the main wine carbonyl compounds. Adapted from Flamini *et al.*, 2005 [46].

Sample volume	100 μL
Vial volume	4 mL
Derivatization conditions	200 μL IS *o*-chlorobenzaldehyde, 3.4 mg/L in ethanol/water solution; 1 mL of PFBOA 2 g/L aqueous solution, volume adjusted to 2 mL with water
SPME fiber	65-μm poly(ethylene glycol)/divinylbenzene (PEG/DVB)
Addition to the sample	50 mg NaCl
Sample heating	50 °C for 20 min under stirring
Extraction temperature and time	50 °C for 5 min
Desorption temperature and time	240 °C for 1 min
Fiber cleaning	250 °C for 5 min
GC column	HP-INNowax (30 m × 0.25 mm i.d; 0.25-μm film thickness)
Carrier gas	Helium, column headpressure 16 psi
Injector	T = 240 °C, sample volume 0.5 μL, splitless injection
Oven program	60 °C for 5 min, 3 °C/min to 210 °C, held 5 min
MS-IT conditions	PICI mode using methane as reagent gas (flow 1 mL/min), ion source at 200 °C, damping gas 0.3 mL/min, simultaneous SCAN (range *m/z* 40–660, 1.67 scan/s) and MS/MS
CID experiments	Collision gas He, excitation voltage 225 mV
Quantitative	Recorded signals at *m/z* 240 for acetaldehyde, *m/z* 282 for diacetyl, *m/z* 266 for acetoin, *m/z* 336 for *o*-chlorobenzaldehyde (I.S.)

In making barrels for wine aging, oak (*Quercus sessilis, Q. petraea, Q. robur, Q. peduncolata, Q. alba*) is the wood more often used but other species, such as acacia (*Robinia pseudoacacia*), chestnut (*Castanea sativa*), cherry (*Prunus avium*), and mulberry (*Morus alba* and *Morus nigra*) are also being considered [55].

SPME-GC/MS was used to study the evolution of wine aroma during aging in 225-L barrels (barriques) made with these wood types. Experimental conditions used are reported in Table 3; main compounds identified are reported in Table 4 [55].

Table 3. SPME-GC/MS conditions used to study the evolution of volatile compounds of Raboso Piave wine during aging in five different types of wood barrels [55].

SPME fiber	65-μm carbowax/divinylbenzene (CAR/DVB)
Sample volume	10 mL
Vial volume	20 mL
Addition to the sample	3 g NaCl
Sample heating	70 °C for 10 min
Extraction temperature and time	70 °C for 30 min
Desorption temperature and time	230 °C fo 5 min
Fiber cleaning	10 min
GC column	HP-INNowax (30 m × 0.25 mm i.d; 0.25 μm film thickness)
Injection	Splitless
Oven program	40 °C for 5 min, 3 °C/min to 230 °C, held 10 min
MS conditions	ionization energy 70 eV, acquisition SIM mode

Wines aged in acacia, chestnut and oak wood showed higher contents of vanillin and eugenol and the acacia-aged sample showed an increase of 4-ethylguaiacol. Mulberry-aged wine had a significant decrease of 4-ethylguaiacol and increase of 4-ethylphenol; the wine aged in cherry barrel already showed high levels of 4-ethylguaiacol after three months of aging.

2.3. "Foxy Smelling Compounds" and 3-Alkyl-2-Methoxypyrazines in Grape Juice

2'-Aminoacetophenone (*o*-AAP) is the main compound identified as the cause of the aging note—the so-called "hybrid note", "foxy-smelling" or

"American character"—typical of *V. labruscana* grapes, even though it was also found in some *V. vinifera* wines such as Müller-Thurgau, Riesling, and Silvaner [56]. This note is variously described as "acacia blossom," "naphthalene note," "furniture polish," "fusel alcohol," and "damp cloth," and causes a considerable number of wine rejections. The formation of *o*-AAP in grape is promoted by several factors, such as reduced nitrogen fertilization in combination with hot and dry summers, and the risk increases in wines made with grapes harvested early. The phytohormone indole-3-acetic acid (IAA) is the principal precursor of *o*-AAP through non-enzymatic processes [57,58]. Also, methyl anthranilate (MA) contributes to the typical foxy taint of wines made with American and wild vine grapes, although it was also found in some *V. vinifera* white wines in concentrations of up to 0.3 μg/L [59].

For analysis of *o*-AAP in wine a direct-immersion SPME method by using a DVB/CAR/PDMS fiber and GC/MS, was proposed [60]; instead, analysis of MA in grape juice was performed by using a PDMS fiber [61].

3-Alkyl-2-methoxypyrazines are present in the grape skin, pulp, and bunch stems. These compounds contribute to the aroma of wines by conferring vegetative, herbaceous, bell pepper, or earthy notes. They are characterized by very low sensory thresholds: for 3-isobutyl-2-methoxypyrazine (IBMP), 3-*sec*-butyl-2-methoxypyrazine (SBMP),and 3-isopropyl-2-methoxypyrazine (IPMP) are between 1 and 2 ng/L in water. The level of IBMP in wine may be 10 times the sensory threshold, whereas SBMP and IPMP are normally comparable to their sensory thresholds [62,63,64,65,66,67,68].

In general, the climate influences the biosynthesis of methoxypyrazines, and higher contents were found in grapes from cooler regions. For example, it was observed that 3-isobutyl-2-methoxypyrazine decreases dramatically during ripening. Methoxypyrazines may be also influenced by the light exposure: in general berries exposed to more sunlight have lower contents [67].

Table 4. Principal compounds studied in Raboso Piave wines aged nine months in 225 L barrels of acacia, cherry, chestnut, mulberry, and oak (nd: not detected; tr: trace (<0.01 ppm) [55]).

Barrel	Months of Aging	Compounds mg/L					
		Furfural	5-Methylfurfural	4-Ethylguaiacol	Eugenol	4-Ethylphenol	Vanillin
Acacia	3	0.02 ± 0.01	0.03 ± 0.01	2.24 ± 0.21	0.009 ± 0.001	0.67 ± 0.07	0.09 ± 0.03
	6	0.04 ± 0.01	0.03 ± 0.01	2.94 ± 0.14	0.015 ± 0.001	0.92 ± 0.08	0.16 ± 0.01
	9	0.03 ± 0.01	0.03 ± 0.01	3.25 ± 0.67	0.021 ± 0.005	1.29 ± 0.41	0.31 ± 0.07
Cherry	3	nd	nd	3.01 ± 1.13	0.008 ± 0.004	1.00 ± 0.44	0.08 ± 0.04
	6	tr	nd	3.13 ± 0.26	0.009 ± 0.001	1.04 ± 0.06	0.10 ± 0.01
	9	nd	nd	2.79 ± 0.51	0.007 ± 0.001	0.86 ± 0.18	0.12 ± 0.03
Chestnut	3	0.04 ± 0.02	0.03 ± 0.01	2.53 ± 0.43	0.024 ± 0.004	0.84 ± 0.20	0.45 ± 0.06
	6	0.04 ± 0.01	0.02 ± 0.02	2.30 ± 0.12	0.035 ± 0.003	0.74 ± 0.08	0.60 ± 0.02
	9	0.07 ± 0.01	0.04 ± 0.01	1.84 ± 0.18	0.026 ± 0.002	0.64 ± 0.04	0.43 ± 0.03
Mulberry	3	tr	nd	2.69 ± 0.75	0.004 ± 0.001	1.06 ± 0.26	0.09 ± 0.03
	6	tr	nd	2.72 ± 0.44	0.006 ± 0.001	1.27 ± 0.26	0.08 ± 0.02
	9	tr	tr	1.84 ± 0.20	0.006 ± 0.001	1.19 ± 0.07	0.08 ± 0.01
Oak	3	0.18 ± 0.08	0.14 ± 0.04	2.51 ± 0.14	0.009 ± 0.001	0.90 ± 0.07	0.27 ± 0.04
	6	0.56 ± 0.16	0.19 ± 0.05	2.08 ± 0.02	0.012 ± 0.003	0.75 ± 0.05	0.34 ± 0.08
	9	0.60 ± 0.06	0.32 ± 0.04	2.90 ± 0.75	0.018 ± 0.005	1.06 ± 0.36	0.36 ± 0.09

SPME-GC/MS of 3-alkyl-2-methoxypyrazines in grape juice and wine was performed by using DVB/CAR/PDMS, PDMS/DVB and CAR/PDMS fibers [69,70,71,72]. Also, a SPME-GC/MS and multiple mass spectrometry (MS/MS) method for simultaneous determination of *o*-AAP, MA, and the main four 3-alkyl-2-methoxypyrazines [ethylmethoxypyrazine (ETMP), IPMP, SBMP, and IBMP] in grape juice was proposed [73]. Quantitation of methoxypyrazines was performed on the signal area of MS/MS ions at *m/z* 119 (for ETMP), *m/z* 109 (IPMP), *m/z* 81 (IBMP), *m/z* 81 (SBMP) using 2-ethoxy-3-isopropylpyrazine as internal standard. For quantification of *o*-AAP and MA, as internal standard 2,4-dichloroaniline was used.

The optimized HS-SPME experimental conditions are described in Table 5 and GC/MS conditions in Table 6 [73].

Table 5. HS-SPME conditions used for simultaneous analysis of "foxy smelling compounds" (*o*-AAP and MA) and 3-alkyl-2-methoxypyrazines (ETMP, IPMP, IBMP, and SBMP) in grape juice [73].

SPME fiber	50/30 μm divinylbenzene/Carboxen™/polydimethylsiloxane (DVB/CAR/PDMS)
Sample volume	10 mL
Vial volume	20 mL
Addition to the sample	3 g NaCl
Extraction temperature and time	50 °C for 30 min
Desorption temperature and time	250 °C fo 5 min
Fiber cleaning	10 min

Table 6. GC/MS conditions used for simultaneous analysis of *o*-AAP, MA and 3-alkyl-2-methoxypyrazines in grape juice [73]. MW: molecular weight.

GC column	HP-5ms: (5%-phenyl) methylpolysiloxane (30 m × 0.25mm i.d; 0.25-μm film thickness)
Carrier gas	Helium at constant flow 1.2 mL/min
Injector	250 °C
Oven program	40 °C for 5 min, 5 °C/min to 230 °C, held 3 min
MSD conditions	Ionization energy 70 eV, transfer line temperature 280 °C, ion source 250 °C, ion trap in MS/MS mode

Analyte	MW	GC retention time (min)	IT-MS/MS Precursor ion *m/z*	IT-MS/MS MS/MS signal *m/z*
3-ethyl-2-methoxypyrazine	138.17	15.10	138	119
3-isopropyl-2-methoxypyrazine	152.20	16.46	137	109
3-isobutyl-2-methoxypyrazine	166.22	18.90	124	81
3-sec-butyl-2-methoxypyrazine	166.22	19.14	138	81
2-ethoxy-3-isopropylpyrazine (IS)	166.22	18.45	166	123
methyl anthranilate	151.16	23.71	151	TIC
2'-aminoacetophenone	135.16	22.59	135	TIC
2,4-dichloroaniline (IS)	162.02	23.35	161	TIC

2.4. Volatile Phenols In Wine

4-Ethylphenol (4-EP) and 4-ethylguaiacol (4-EG) are associated with wine defects that can form during winemaking or, more commonly, wine aging. These compounds are characterized by sensorial characteristics described as "stable", "animal" and "phenolic" and their presence is particularly detrimental for the product [74,75,76]. They are produced by winery contaminants, such as *Brettanomyces* and *Dekkera* yeasts, through processes of decarboxylation and reduction of ferulic and *p*-coumaric acids present in the grape [75]. Sensory thresholds of 4-EP and 4-EG in wine are 440 μg/L and 33 μg/L, respectively [77].

For their determination several sensitive SPME-GC/MS methods were proposed [20,78,79]. Martorell *et al.*, proposed the use of two PDMS 100 μm fibers: for 4-EP the method had a limit of detection (LOD) and of quantification (LOQ) of 2 μg/L and 5 μg/L, respectively, for 4-EG 1 μg/L and 5 μg/L, respectively [80]. An optimized method for analysis of ethylphenols and vinylphenols in white and red wines was developed by the use of StableFlex Carbowax/DVB (CW/DVB 70 μm) and polyacrilate (PA 85 μm) fibers [79].

4-EG and 4-EP in wine were also analyzed by a multiple-headspace SPME method. By performing three consecutive extractions of the sample with a CW/DVB fiber, the possible matrix effects were minimized by providing a LOD of 0.06 μg/L for both 4-EG and 4-EP [81].

2.5. Higher Alcohols and Esters in Wine

By using a PDMS 100-μm fiber, effective methods for analysis of higher alcohols and aliphatic esters in wine were performed [82,83]. This coating fiber showed high affinity for non-polar compounds such as ethyl esters and acetates [84,85,86], while CW/DVB fiber is suitable for more polar compounds, such as 1-hexanol, hexen-1-ol, 1-octanol, and monoterpenols [84].

Antalick *et al.*, developed a SPME and GC/MS-SIM method which provided simultaneous determination of 32 esters in wine in concentration between ng/L and mg/L [87]. Seven different fibers were tested: DVB/CAR/PDMS 50/30 µm, CAR/PDMS 85 µm, PDMS 100 µm, PDMS/DVB 65 µm, PA 85 µm, CW/DVB 70 µm, and polyethyleneglycol (PEG) 60 µm. PDMS was the most efficient in extracting the less polar and less volatile compounds; for more volatile esters the best coating was CAR/PDMS, and aromatic esters were better recovered by CW/DVB. In general, the PDMS fiber showed high efficiency for all compounds and provided LOQs between 0.4 ng/L and 4.0 µg/L.

Recent applications showed that the tri-phase fiber DVB/CAR/PDMS provides extraction of the highest number of wine volatiles, including ethyl esters (56% of the compounds identified), alcohols, and acids [2,88].

2.6. Wine Volatile Sulfur Compounds

Various sulfur compounds are present in wine, such as thiols, sulphides, thioesters, and heterocyclic compounds. Thiol and thio-type compounds are in general associated with flavor defects of wine and they are classified as "light" (boiling point < 90 °C) and "heavy" (b.p. > 90 °C) compounds [25,29,89,90,91]. Sulfur compounds can be formed through several enzymatic and non-enzymatic processes, such as yeast fermentation and chemical, photochemical, and thermal reactions occurring in winemaking and during wine storage [89,90]. Most prevailing are ethylmercaptan (EtSH; onion as aroma descriptor), dimethyl sulfide (DMS; grassy/truffle-like note), 2-furanmethanethiol (FFT; roasted coffee), diethyl sulfide (DES; cooked vegetables, onion, garlic), dimethyl disulfide (DMDS; cooked gabbage, intense onion), diethyl disulfide (DEDS; garlic, burnt rubber), methyl thioacetate (MTA), ethyl thioacetate (ETA), 2-mercaptoethanol (ME; burnt rubber), 2-(methylthio)-1-ethanol (MTE; cauliflower), 3-(methylthio)-1-propanol (MTP; sweet, potato), 4-(methylthio)-1-butanol (MTB; earthy-like scent), benzothiazole (BT; rubber), and 5-(2-hydroxyethyl)-4-methylthiazole (HMT) [29,91]. Also, 2-methyl-3-furanthiol (MF; cooked meat), a very odoriferous compound with an odor threshold of 0.4–1.0 ppt, was found [92].

A SPME method for analysis of 13 volatile sulfur compounds in wine (*i.e.*, DMS, EtSH, DES, MTA, ETA, ME, DMDS, DEDS, BT, HMT, MTB, MTP, and MTE) with b.p. ranging from 35 °C to 231 °C by using a CAR/PDMS/DVB 50:30 μm 2 cm length fiber, was developed [90]. By addition of $MgSO_4$ (1.0 M) to increase ionic strength of solution and performing the extraction at 35 °C, the method showed a high sensitivity for all the analytes.

3-mercaptohexan-1-ol (3-MH; passion fruit, grapefruit), 3-mercaptohexyl acetate (3-MHA; Riesling type note, passion fruit, box tree), and 4-methyl-4-mercaptopentan-2-one (4–MP; blackcurrant or box tree note) are tropical fruit scenting volatiles present in wines at ng/L level [29]. A SPME-GC/MS method for analysis of these compounds by using a CAR/PDMS/DVB fiber was proposed [93]. HS-SPME conditions were optimized by performing the extraction of wine adjusted to pH 7 at 40 °C for 40 min [93].

A method for analysis of thiols in wine by synthesis of the pentafluorobenzyl derivatives was also proposed. Derivatization was performed directly on the PDMS/DVB fiber (65 μm), LODs achieved were 0.05 ng/L, 0.03 ng/L, 0.11 ng/L, 0.5 ng/L, and 0.8 ng/L for FFT, 3-MHA, MF, 4-MP, and 3-MH, respectively [94].

3. CONCLUSIONS

Grape aroma is composed of a hundred compounds and wine's volatile profile also includes a number of fermentative compounds. SPME coupled to GC/MS showed to be effective for studying several classes of these analytes without solvent use. It often resulted in a high-sensitive technique for quantitative analysis of compounds for which the standard is available with high reproducibility. Moreover, the use of a multiphase fiber coupled to MS and multivariate data analysis allows sampling automation and statistical treatment of fragment abundances for the identification of compounds [95,96].

On the other hand, different to most of the sample preparation methods performed by liquid-liquid extraction and SPE, the selectivity of SPME fiber often changes dramatically for the different analytes. As a consequence, it is rarely possible to perform the semi-quantitative profiling of the sample on the internal standard signal, which is particularly useful in the characterization of grape varieties and the monitoring of the winemaking processes.

Logically, by increasing the standards commercially available new SPME-GC/MS applications are developed, and methods for the profiling of specific classes of grape aroma compounds, such as terpenols and norisoprenoids, could be particularly useful.

AUTHOR CONTRIBUTIONS

A.P. and R.F. wrote the paper.

REFERENCES

1 Weldegergis, B.T.; Croucha, A.M.; Górecki, T.; de Villiers, A. Solid phase extraction in combination with comprehensive two-dimensional gas chromatography coupled to time-of-flight mass spectrometry for the detailed investigation of volatiles in South African red wines. *Anal. Chim. Acta* **2011**, *701*, 98–111.

2 Sagratini, G.; Maggi, F.; Caprioli, G.; Cristalli, G.; Ricciutelli, M.; Torregiani, E.; Vittori, S. Comparative study of aroma profile and phenolic content of Montepulciano monovarietal red wines from the Marches and Abruzzo regions of Italy using HS-SPME-GC-MS and HPLC-MS. *Food Chem.* **2012**, *132*, 1592–1599.

3 Blanch, G.P.; Reglero, G.; Herraiz, M. Rapid extraction of wine aroma compounds using a new simultaneous distillation-solvent extraction device. *Food Chem.* **1996**, *56*, 439–444.

4 Bosch-Fusté, J.; Riu-Aumatell, M.; Guadayol, J.M.; Caixach, J.; López-Tamames, E.; Buxaderas, S. Volatile profiles of sparkling wines obtained by three extraction methods and gas chromatography-mass spectrometry (GC-MS) analysis.*Food Chem.* **2007**, *105*, 428–435.

5 Mayr, C.M.; Geue, J.P.; Holt, H.E.; Pearson, W.P.; Jeffery, D.W.; Francis, I.L. Characterization of the key aroma compounds in Shiraz wine by quantitation, aroma reconstitution, and omission studies. *J. Agric. Food Chem.* **2014**, *62*, 4528–4536.

6 Mamede, M.; Pastore, G.M. Study of methods for the extraction of volatile compounds from fermented grape must.*Food Chem.* **2006**, *96*, 586–590.

7 Mateo, J.J.; Gentilini, N.; Huerta, T.; Jiménez, M.; di Stefano, R. Fractionation of glycoside precursors of aroma in grape and wine. *J. Chromatogr. A* **1997**, *778*, 219–224.

8 Cabrita, M.J.; Costa Freitas, A.M.; Laureano, O.; Borsa, D.; di Stefano, R. Aroma compounds in varietal wines from Alentejo, Portugal. *J. Food Comp. Anal.* **2007**, *20*, 375–390.

9 Rosillo, L.; Salinas, M.R.; Garijo, J.; Alonso, G.L. Study of volatiles in grapes by dynamic headspace analysis: Application to the differentiation of some *Vitis vinifera* varieties. *J. Chromatogr. A* **1999**, *847*, 155–159.

10 Marengo, E.; Aceto, M.; Maurino, V. Classification of Nebbiolo-based wines from Piedmont (Italy) by means of solid-phase microextraction-gaschromatography-mass spectrometry of volatile compounds. *J. Chromatogr. A* **2001**, *943*, 123–137.

11 Sanchéz-Palomo, E.; Díaz-Maroto, M.C.; Pérez-Coello, S. Rapid determination of volatile compounds in grapes by HS-SPME coupled with GC-MS. *Talanta* **2005**, *66*, 1152–1157.

12 Coelho, E.; Rocha, S.M.; Delgadillo, I.; Coimbra, M.A. Headspace-SPME applied to varietal volatile components evolution during *Vitis vinifera* L. cv. "Baga" ripening. *Anal. Chim. Acta* **2006**, *563*, 204–214.

13 Verzera, A.; Ziino, M.; Scacco, A.; Lanza, C.M.; Mazzaglia, A.; Romeo, V.; Condurso, C. Volatile compound and sensory analysis for the characterization of an Italian white wine from "Inzolia" grapes. *Food Anal. Methods* **2008**, *1*, 144–151.

14 Arthur, C.L.; Pawliszyn, J. Solid-phase microextraction with thermal desorption using silica optical fibers. *Anal. Chem.* **1990**, *62*, 2145–2148.

15 Bojko, B.; Cudjoe, E.; Gómez-Ríos, G.A.; Gorynski, K.; Jiang, R.; Reyes-Garcés, N.; Risticevic, S.; Silva, É.A.S.; Togunde, O.; Vuckovic, D.; *et al.* SPME-Quo vadis? *Anal. Chim. Acta* **2012**, *750*, 132–151.

16 Harmon, A.D. Solid-phase microextraction for the analysis of flavors. In *Techniques for Analyzing Food Aroma*; Marsili, R., Ed.; Marcel Decker, Inc.: New York, NY, USA, 1997; pp. 81–112.

17 Yu, Y.-J.; Lu, Z.-M.; Yu, N.-H.; Xu, W.; Li, G.-Q.; Shi, J.-S.; Xu, Z.-H. HS-SPME/GC-MS and chemometrics for volatile composition of Chinese traditional aromatic vinegar in the Zhenjiang region. *J. Inst. Brew.* **2012**, *118*, 133–141.

18 Vas, G.; Vékey, K. Solid-phase microextraction: A powerful sample preparation tool prior to mass spectrometric analysis. *J. Mass Spectrom.* **2004**, *39*, 233–254.

19 Castro, R.; Natera, R.; Durán, E.; García-Barroso, C. Application of solid phase extraction techniques to analyse volatile compounds in wines and other enological products. *Eur. Food Res. Technol.* **2008**, *228*, 1–18.

20 Flamini, R. Volatile and aroma compounds in wines. In *Mass Spectrometry in Grape and Wine Chemistry*; Flamini, R., Traldi, P., Eds.; John Wiley & Sons, Inc.: Hoboken, NJ, USA, 2010; pp. 117–162.

21 Garcóa, E; Chacón, J.L.; Martínez, J.; Izquierdo, P.M. Changes in volatile compounds during ripening in grapes of Airén, Macabeo and Chardonnay white varieties grown in La Mancha Region (Spain). *Food Sci. Technol. Int.* **2003**, *9*, 33–41.

22 Noguerol-Pato, R.; González-Barreiro, C.; Cancho-Grande, B.; Santiago, J.L.; Martínez, M.C.; Simal-Gándara, J. Aroma potential of Brancellao grapes from different cluster positions. *Food Chem.* **2012**, *132*, 112–124.

23 Marais, J. Terpenes in the aroma of grapes and wines: A review. *S. Afr. J. Enol. Vitic.* **1983**, *4*, 49–58.

24 Flamini, R. Grape aroma compounds: Terpenes, C_{13}-norisoprenoids, benzene compounds, and 3-alkyl-2-methoxypyrazines. In *Mass Spectrometry in Grape and Wine Chemistry*; Flamini, R., Traldi, P., Eds.; John Wiley & Sons, Inc.: Hoboken, NJ, USA, 2010; pp. 97–116.

25 Riberéau-Gayon, P.; Glories, Y.; Maujean, A.; Dubourdieu, D. Varietal Aroma. In *Handbook of Enology: The Chemistry of Wine. Stabilization and Treatments*, 2nd ed.; John Wiley & Sons Ltd: Chichester, UK, 2006; Volume 2, pp. 205–230.

26 Wang, Y.; Kays, S.J. Contribution of volatile compounds to the characteristic aroma of baked Jewel sweetpotatos. *J. Am. Soc. Hortic. Sci.* **2000**, *125*, 638–643.

27 Robinson, A.L.; Boss, P.K.; Solomon, P.S.; Trengove, R.D.; Heymann, H.; Ebeler, S.E. Origins of grape and wine aroma. Part 1. Chemical Components and Viticultural Impacts. *Am. J. Enol. Vitic.* **2014**, *65*, 1–24.

28 Riberéau-Gayon, P.; Glories, Y.; Maujean, A.; Dubourdieu, D. Alcohols and other volatile compounds. In *Handbook of Enology: The Chemistry of Wine. Stabilization and Treatments*, 2nd ed.; John Wiley & Sons Ltd: Chichester, UK, 2006; Volume 2, pp. 51–64.

29 Bakker, J.; Clarke, R.J. Volatile components. In *Wine Flavour Chemistry*, 2nd ed.; John Wiley & Sons, Ltd: Chichester, UK, 2012; pp. 155–238.

30 Flamini, R.; de Luca, G.; di Stefano, R. Changes in carbonyl compounds in Chardonnay and Cabernet Sauvignon wines as a consequence of malolactic fermentation. *Vitis* **2002**, *41*, 107–112.

31 Flamini, R.; Dalla Vedova, A. Glyoxal/glycolaldehyde: A redox system involved in malolactic fermentation of wine. *J. Agric. Food Chem.* **2003**, *51*, 2300–2303.

32 Baumes, R. Wine Aroma Precursors. In *Wine Chemistry and Biochemistry*; Victoria Moreno-Arribas, M., Carmen Polo, M., Eds.; Springer Science + Business Media, LLC: New York, NY, USA, 2009; pp. 251–274.

33 Gambetta, J.M.; Bastian, S.E. P.; Cozzolino, D.; Jeffery, D.W. Factors influencing the aroma composition of Chardonnay wines. *J. Agric. Food Chem.* **2014**, *62*, 6512–6534.

34 Bell, S.J.; Henschke, P.A. Implications of nitrogen nutrition for grapes, fermentation and wine. *Aust. J. Grape Wine Res.***2005**, *11*, 242–295.

35 Swiegers, J.H.; Bartowsky, E.J; Henschke, P.A.; Pretorius, I.S. Yeast and bacterial modulation of wine aroma and flavour. *Aust. J. Grape Wine Res.* **2005**, *11*, 139–173.

36 Rapp, A.; Versini, G. Influence of nitrogen compounds in grapes on aroma compounds in wine. In Proceedings of the International Symposium on Nitrogen in Grapes and Wine, Seattle, DC, USA, 18–19 June 1991; American Society of Enology and Viticulture: Davis, CA, USA, 1991; pp. 156–164.

37 Riberéau-Gayon, P.; Glories, Y.; Maujean, A.; Dubourdieu, D. Phenolic compounds. In *Handbook of Enology: The Chemistry of*

Wine. Stabilization and Treatments, 2nd ed.; John Wiley & Sons Ltd: Chichester, UK, 2006; Volume 2, pp. 141–203.

38 Cancilla, D.A.; Que Hee, S.S. *O*-(2,3,4,5,6-Pentafluorophenyl)methylhydroxylamine hydrochloride: A versatile reagent for the determination of carbonyl-containing compounds. *J. Chromatogr. A* **1992**, *627*, 1–16.

39 Vanderlinde, R.; Bertrand, A.; Segur, M.C. Dosage des aldehydes dans les eaux-de-vie. In Proceedings of the 1er Symposium Scientifique International du Cognac, "Elaboration et connaissance des spiritueux", Cognac, France, 11–15 May 1992; Lavoiser TEC & DOC: Paris, France, 1992; pp. 506–511.

40 Vidal, J.P.; Mazerolles, G.; Estreguil, S.; Cantagrel, R. Analyse quantitative de la fraction carbonylée volatile des eaux-de-vie de Cognac. In Proceedings of the 1er Symposium Scientifique International du Cognac, "Elaboration et connaissance des spiritueux", Cognac, France, 11–15 May 1992; Lavoiser TEC & DOC: Paris, France, 1992; pp. 529–537.

41 De Revel, G.; Bertrand, A. A method for the detection of carbonyl compounds in wine: Glyoxal and methylglyoxal. *J. Sci. Food Agric.* **1993**, *61*, 267–272.

42 De Revel, G.; Bertrand, A. Dicarbonyl compounds and their reduction products in wine. Identification of wine aldehydes. In *Trends in Flavour Research*; Maarse, H., van der Heij, D.G., Eds.; Elsevier Science B.V.: Amsterdam, The Netherlands, 1994; pp. 353–361.

43 Guillou, I.; Bertrand, A.; de Revel, G.; Barbe, J.C. Occurrence of hydroxypropanedial in certain musts and wines. *J. Agric. Food Chem.* **1997**, *45*, 3382–3386.

44 Flamini, R.; Tonus, T.; Dalla Vedova, A. A GC-MS method for determining acetaldehyde in wines. *Riv. Vitic. Enol.***2002**, *2/3*, 15–21.

45 Flamini, R.; Dalla Vedova, A.; Panighel, A. Study of carbonyl compounds in some Italian marc distillate (grappa) samples by synthesis of *O*-(2,3,4,5,6-pentafluorobenzyl)-hydroxylamine derivatives. *Riv. Vitic. Enol.* **2005**, *1*, 51–63.

46 Flamini, R.; Dalla Vedova, A.; Panighel, A.; Perchiazzi, N.; Ongarato, S. Monitoring of the principal carbonyl compounds involved in malolactic fermentation of wine by solid-phase microextraction and positive ion chemical ionization GC/MS analysis. *J. Mass Spectrom.* **2005**, *40*, 1558–1564.

47 López-Vázquez, C.; Orriols, I.; Perelló, M.-C.; de Revel, G. Determination of aldehydes as pentafluorobenzyl derivatives in grape pomace distillates by HS-SPME-GC/MS. *Food Chem.* **2012**, *130*, 1127–1133.

48 Bauer, R.; Dicks, L.M.T. Control of Malolactic Fermentation in Wine. A Review. *S. Afr. J. Enol. Vitic.* **2004**, *25*, 74–88.

49 Lerm, E.; Engelbrecht, L.; du Toit, M. Malolactic Fermentation: The ABC's of MLF. *S. Afr. J. Enol. Vitic.* **2010**, *31*, 186–212.

50 Zapata, J.; Mateo-Vivaracho, L.; Cacho, J.; Ferreira, V. Comparison of extraction techniques and mass spectrometric ionization modes in the analysis of wine volatile carbonyls. *Anal. Chim. Acta* **2010**, *660*, 197–205.

51 Pérez-Coello, M.S.; Sanz, J.; Cabezudo, M.D. Determination of volatile compounds in hydroalcoholic extracts of French and American oak wood. *Am. J. Enol. Vitic.* **1999**, *50*, 162–165.

52 Garde-Cedàn, T.; Lorenzo, C.; Carot, J.M.; Jabaloyes, J.M.; Esteve, M.D.; Salinas, M.R. Statistical differentiation of wines of different geographic origin and aged in barrel according to some volatile components and ethylphenols. *Food Chem.* **2008**, *111*, 1025–1031.

53 Dìaz-Maroto, M.C.; Sànchez-Palomo, E.; Pérez-Coello, M.S. Fast screening method for volatile compounds of oak wood used for aging

wines by headspace SPME-GC-MS (SIM). *J. Agric. Food Chem.* **2004**, *52*, 6857–6861.

54 Carillo, J.D.; Garrido-Lòpez, A.; Tena, M.T. Determination of oak volatile compounds in wine by headspace solid-phase microextraction and gas chromatography-mass spectrometry. *J. Chromatogr. A* **2006**, *1102*, 25–36.

55 De Rosso, M.; Panighel, A.; Dalla Vedova, A.; Stella, L.; Flamini, R. Changes in chemical composition of a red wine aged in acacia, cherry, chestnut, mulberry, and oak wood barrels. *J. Agric. Food Chem.* **2009**, *57*, 1915–1920.

56 Rapp, A.; Versini, G.; Ullemeyer, H. 2-aminoacetophenone: causal component of "untypical aging flavor" ("naphtalene note", "hybrid note") of wine. *Vitis* **1993**, *32*, 61–62.

57 Hoenicke, K.; Borchert, O.; Grüning, K.; Simat, T.J. "Untypical aging off-flavor" in wine: synthesis of potential degradation compounds of indole-3-acetic acid and kynurenine and their evaluation as precursors of 2-aminoacetophenone. *J. Agric. Food Chem.* **2002**, *50*, 4303–4309.

58 Hoenicke, K.; Simat, T.J.; Steinhart, H.; Christoph, N.; Geßner, M.; Köhler, H.-J. "Untypical aging off-flavor" in wine: formation of 2-aminoacetophenone and evaluation of its influencing factors. *Anal. Chim. Acta* **2002**, *458*, 29–37.

59 Rapp, A.; Versini, G. Methylanthranilate ("foxy taint") concentrations of hybrid and *Vitis vinifera* wines. *Vitis* **1996**, *35*, 215–216.

60 Fan, W.; Tsai, I.-M.; Qian, M.C. Analysis of 2-aminoacetophenone by direct-immersion solid-phase microextraction and gas chromatography-mass spectrometry and its sensory impact in Chardonnay and Pinot gris wines. *Food Chem.***2007**, *105*, 1144–1150.

61 Massa, M.J.; Robacker, D.C.; Patt, J. Identification of grape juice aroma volatiles and attractiveness to the Mexican fruit fly (*Diptera: Tephritidae*). *Florida Entomologist* **2008**, *91*, 266–276.

62 Lacey, M.; Allen, M.S.; Harris, R.L.N.; Brown, W.V. Methoxypyrazines in Sauvignon blanc grapes and wines. *Am. J. Enol. Vitic.* **1991**, *42*, 103–108.

63 Allen, M.S.; Lacey, M.J.; Harris, R.L.N.; Brown, W.V. Contribution of Methoxypyrazines to Sauvignon blanc Wine Aroma. *Am. J. Enol. Vitic.* **1991**, *42*, 109–112.

64 Allen, M.S.; Lacey, M.J. Methoxypyrazine grape flavour: Influence of climate, cultivar and viticulture. *Die Wein-Wiss.***1993**, *48*, 211–213.

65 Hashizume, K.; Umeda, N. Methoxypyrazine content of Japanese red wines. *Biosci. Biotechnol. Biochem.* **1996**, *60*, 802–805.

66 Hashizume, K.; Samuta, T.J. Green odorants of grape cluster stem and their ability to cause a wine stemmy flavor. *J. Agric. Food Chem.* **1997**, *45*, 1333–1337.

67 Hashizume, K.; Samuta, T. Grape maturity and light exposure affect berry methoxypyrazine concentration. *Am. J. Enol. Vitic.* **1999**, *50*, 194–198.

68 Roujou de Boubee, D.; Cumsille, A.M.; Pons, D.; Dubordieu, D. Location of 2-methoxy-3-isobutylpyrazine in Cabernet Sauvignon grape bunches and its extractability during vinification. *Am. J. Enol. Vitic.* **2002**, *53*, 1–5.

69 Kotseridis, Y.S.; Spink, M.; Brindle, I.D.; Blake, A.J.; Sears, M.; Chen, X.; Soleas, G.; Inglis, D.; Pickering, G.J. Quantitative analysis of 3-alkyl-2-methoxypyrazines in juice and wine using stable isotope labelled internal standard assay. *J. Chromatogr. A* **2008**, *1190*, 294–301.

70 Sala, C.; Mestres, M.; Martì, M.P.; Busto, O.; Guasch, J. Headspace solid-phase microextraction analysis of 3-alkyl-2- methoxypyrazines in wines. *J. Chromatogr. A* **2002**, *953*, 1–6.

71 Galvan, T.L.; Kells, S.; Hutchison, W.D. Determination of 3-alkyl-2-methoxypyrazines in Lady beetle-Infested wine by Solid-Phase Microextraction headspace sampling. *J. Agric. Food Chem.* **2008**, *56*, 1065–1071.

72 Ryan, D.; Watkins, P.; Smith, J.; Allen, M.; Marriott, P. Analysis of methoxypyrazines in wine using headspace solid phase microextraction with isotope dilution and comprehensive two-dimensional gas chromatography. *J. Sep. Sci.***2005**, *28*, 1075–1082.

73 Panighel, A.; Dalla Vedova, A.; de Rosso, M.; Gardiman, M.; Flamini, R. A solid-phase microextraction gas chromatography/ion trap tandem mass spectrometry method for simultaneous determination of "foxy smelling compounds" and 3-alkyl-2-methoxypyrazines in grape juice. *Rapid Commun. Mass Spectrom.* **2010**, *24*, 2023–2029.

74 Chatonnet, P.; Boidron, J.-N.; Pons, M. Elevage des vins rouges en fûts de chêne: Évolution de certains composés volatils et de leur impact aromatique. *Sci. Aliment.* **1990**, *10*, 565–587.

75 Chatonnet, P.; Dubourdieu, D.; Boidron, J.-N.; Pons, M. The origin of ethylphenols in wines. *J. Sci. Food Agric.* **1992**, *60*, 165–178.

76 Rodrigues, N.; Gonçalves, G.; Pereira-da-Silva, S.; Malfeito-Ferreira, M.; Loureiro, V. Development and use of a new medium to detect yeast of the genera *Dekkera/Brettanomyces* sp. *J. Appl. Microbiol.* **2001**, *90*, 588–599.

77 Carrillo, J.D.; Tena, M.T. Determination of ethylphenols in wine by *in situ* derivatisation and headspace solid-phase microextraction gas chromatography mass spectrometry. *Anal. Bioanal. Chem.* **2007**, *387*, 2547–2558.

78 Monje, M.-C.; Privat, C.; Gastine, V.; Nepveu, F. Determination of ethylphenol compound in wine by headspace solid-phase microextraction in conjunction with gas chromatography and flame ionization detection. *Anal. Chim. Acta* **2002**,*458*, 111–117.

79 Castro Mejías, R.; Natera Marín, R.; García Moreno, M.d.V.; García Barroso, C. Optimisation of headspace solid-phase microextraction for the analysis of volatile phenols in wine. *J. Chromatogr. A* **2003**, *995*, 11–20.

80 Martorell, N.; Martì, M.P.; Mestres, M.; Busto, O.; Guasch, J. Determination of 4-ethylguaiacol and 4-ethylphenol in red wines using headspace-solid-phase microextraction-gas chromatography. *J. Chromatogr. A* **2002**, *975*, 349–354.

81 Pizarro, C.; Pérez-del-Notario, N.; Gonzàlez-Sàiz, J.M. Determination of Brett character responsible compounds in wines by using multiple headspace solid-phase microextraction. *J. Chromatogr. A* **2007**, *1143*, 176–181.

82 Francioli, S.; Guerra, M.; López-Tamames, E.; Guadayoi, J.M.; Caixach, J. Aroma of sparkling wines by headspace/solid phase microextraction and gas chromatography/mass spectrometry. *Am. J. Enol. Vitic.* **1999**, *50*, 404–408.

83 Demyttenaere, J.C.R.; Dagher, C.; Sandra, P.; Kallithraka, S.; Verhé, R.; de Kimpe, N. Flavour analysis of Greek white wine by solid-phase microextraction-capillary gas chromatography-mass spectrometry. *J. Chromatogr. A* **2003**, *985*, 233–246.

84 Bonino, M.; Schellino, R.; Rizzi, C.; Aigotti, R.; Delfini, C.; Baiocchi, C. Aroma compounds of an Italian wine (*Ruchè*) by HS-SPME analysis coupled with GC/IT/MS. *Food Chem.* **2003**, *80*, 125–133.

85 Vianna, E.; Ebeler, S. Monitoring ester formation in grape juice fermentations using solid phase microextraction coupled with gas

chromatography-mass spectrometry. *J. Agric. Food Chem.* **2001**, *49*, 589–595.

86 Martí, M.P.; Mestres, M.; Sala, C.; Busto, O.; Guasch, J. Solid-phase microextraction and gas chromatography olfactometry analysis of successively diluted samples. A new approach of the aroma extract dilution analysis applied to the characterization of wine aroma. *J. Agric. Food Chem.* **2003**, *51*, 7861–7865.

87 Antalick, G.; Perello, M.-C.; de Revel, G. Development, validation and application of a specific method for the quantitative determination of wine esters by headspace-solid-phase microextraction-gas chromatography-mass spectrometry. *Food Chem.* **2010**, *121*, 1236–1245.

88 Welke, J.E.; Zanus, M.; Lazarotto, M.; Schmitta, K.G.; Zini, C.A. Volatile characterization by multivariate optimization of headspace-solid phase microextraction and sensorial evaluation of Chardonnay base wines. *J. Braz. Chem. Soc.* **2012**,*23*, 678–687.

89 Mestres, M.; Busto, O.; Guasch, J. Analysis of organic sulfur compounds in wine aroma. *J. Chromatogr. A* **2000**, *881*, 569–581.

90 Fedrizzi, B.; Magno, F.; Moser, S.; Nicolini, G.; Versini, G. Concurrent quantification of light and heavy sulphur volatiles in wine by headspace solid-phase microextraction coupled with gas chromatography/mass spectrometry.*Rapid Commun. Mass Spectrom.* **2007**, *21*, 707–714.

91 Flamini, R. Analysis of aroma compounds in wine. In *Hyphenated Techniques in Grape and Wine Chemistry*; Flamini, R., Ed.; John Wiley & Sons, Inc.: Hoboken, NJ, USA, 2008; pp. 173–225.

92 Bouchilloux, P.; Darriet, P.; Dubourdieu, D. Identification d'un thiol fortement odorant, le 2-methyl-3-furanthiol, dans les vins. *Vitis* **1998**, *37*, 177–180.

93 Fedrizzi, B.; Versini, G.; Lavagnini, I.; Nicolini, G.; Magno, F. Gas chromatography-mass spectrometry determination of 3-mercaptohexan-1-ol and 3-mercaptohexyl acetate in wine. A comparison between Solid Phase Extraction and Headspace Solid Phase Microextraction methods. *Anal. Chim. Acta* **2007**, *596*, 291–297.

94 Mateo-Vivaracho, L.; Ferreira, V.; Cacho, J. Automated analysis of 2-methyl-3-furanthiol and 3-mercaptohexyl acetate at ng/L level by headspace solid-phase microextracion with on—Fibre derivatisation and gas chromatography-negative chemical ionization mass spectrometric determination. *J. Chromatogr. A* **2006**, *1121*, 1–9.

95 Kinton, V.R.; Collins, R.J.; Kolahgar, B.; Goodner, K.L. Fast analysis of beverages using mass spectral based chemical sensor. *Gerstel AppNote* **2003**, *4*, 1–10.

96 Cozzolino, D.; Smyth, H.E.; Lattey, K.A.; Cynkar, W.; Janik, L.; Dambergs, R.G.; Francis, I.L.; Gishen, M. Combining mass spectrometry based electronic nose, visible-near infrared spectroscopy and chemometrics to assess the sensory properties of Australian Riesling wines. *Anal. Chim. Acta* **2006**, *563*, 319–324.

CHAPTER 2

Multiresidue Method for Determination of 67 Pesticides in Water Samples Using Solid-Phase Extraction with Centrifugation and Gas Chromatography-Mass Spectrometry

Abir Kouzayha[1], Abdul Rahman Rabaa[2], Mohamad Al Iskandarani[1], Daniel Beh[2], Hélène Budzinski[3], Farouk Jaber[1,2]

[1] *Analysis of Pesticides and Organic Pollutants Laboratory LAPPO, Lebanese Atomic Energy Commission LAEC, National Council for Scientific Research CNRS, Beirut, Lebanon*

[2] *Laboratory of Analysis of Organic Compounds (509), Faculty of Sciences I, Lebanese University, Beirut, Lebanon*

[3] *ISM-LPTC-UMR 5255 (Laboratory of Physicoand Toxico-Chemistry), CNRS, Université Bordeaux, Talence, France*

ABSTRACT

A new multi-residue method based on solid-phase extraction (SPE) with centrifugation was developed for determination and quantitation of 67 pesticides in water samples. Two SPE cartridges were tested: Chromabond C_{18} and Oasis HLB. Parameters that influence the extraction efficiency such as the eluent volume, the sample loading volume, the addition of organic solvent to water sample, sorbent drying and elute concentration were optimized. The innovation of this work was the examination of the use of a centrifugation technique in both the drying and elution steps. When combined with centrifugation, the volume of the elution solvent was reduced to 2 mL and the time for sorbent drying decreased also to 10 min under vacuum. Under the optimized conditions, this method showed good recoveries higher than 65% - 68% for the 67 analyzed pesticides using the C_{18} and HLB cartridges with relative standard deviations lower than 9.7% - 12.3%. Limits of quantification were between 2 and 20 ng·L^{-1}. The simplicity of the described method, use of less of organic solvent, short procedure time, and good recoveries demonstrate the advantages of this environmentally friendly approach for routine analysis of numerous samples.

KEYWORDS

Multi-Residue; Pesticides; Solid-Phase Extraction; Centrifugation;
Gas Chromatography-Mass Spectrometry

1. INTRODUCTION

Organic contaminants present in the environment are a result of different sources of pollution from anthropogenic activities [1]. The pesticides, generated by the intensification of agriculture, are regarded as some of the most dangerous contaminants of the environment, despite their numerous merits. Not only are they toxic; they are also mobile and capable of bioaccumulation. On top of this, they can take part in various physical, chemical and biological processes. Many of these pesticides are characterized by a strong persistence which explains their wide presence in the different compartments of the environment [1,2]. Due to these physicochemical characteristics and their extensive use, many of these pesticides end-up in surface and ground water. They are found nowadays in all surface waters and in a growing number of aquifers. Their presence in water is considered as a potential risk not only for drinking water quality and human health, but also for ecosystems [3].

In this context, strict regulations for the control of pesticide residues concentration levels in environment have been established. According to the European Union Directive on drinking water quality (98/83/EC), the admissible concentration for pesticides is 0.1 $\mu g{\cdot}L^{-1}$ for each individual substance and 0.5 $\mu g{\cdot}L^{-1}$ for the sum of pesticides (including their metabolites) [4]. In the field of environmental water policy, the European Union adopted the Water Framework Directive (WFD) (2000/60/EC), whose objectives are to improve, protect and prevent further deterioration of water quality. The maximal concentrations authorized can vary from the $ng{\cdot}L^{-1}$ order to few $\mu g{\cdot}L^{-1}$ [5]. Such imposed regulations have driven the development of novel analytical techniques and improvement of existing ones, so that the largest possible number of compounds at low levels can be determined in one sensitive method.

For instrumental analysis, gas chromatography (GC) or liquid chromatography (LC) followed by mass spectrometry detection (MS) or tandem mass spectrometry (MS-MS) are the most commonly used techniques for the quantification of pesticides in water. However, in spite of the recent technical progress, the pesticide quantification in water requires a first step of extraction and preconcentration [6]. This sample preparation step is, in fact, the critical step of the whole analysis. The most common method is solidphase extraction (SPE) [7-9], which can be used to determine a broad range of pesticides in one analysis.

Modern trends in sample preparation are towards the simplification of the process and minimizing the organic solvent used. In our previous paper [10], we have reported a rapid and reliable method based on SPE with centrifugation for the extraction and followed by GC/MS detection for the determination of the 16 PAH identified by the US environmental protection agency (US-EPA) as priority pollutants in water samples. The focus of the present research is the development of a similar rapid multi-residue method for the analysis and ultra-trace quantification of 67 pesticides from different classes in waters. The potential factors affecting SPE extraction of the analytes were optimized and discussed in details. Then, the whole procedure was validated.

2. EXPERIMENTAL

2.1. Reagents, Standards and Small Apparatus

Acetonitrile (ACN), acetone, dichloromethane (DCM), hexane (Hex) and methanol (MeOH) (HPLC grade ≥ 99.9%) were purchased from Sigma-Aldrich (SigmaAldrich, St Louis, MO, USA). Pesticide standards of high purity level (98% - 99.8%) were purchased from Dr. Ehrenstorfer GmbH (Augsburg, Germany) and ChemService (West Chester, PA, USA). Individual standard pesticide solutions were prepared in acetonitrile with a concentration of about 1000 $mg{\cdot}L^{-1}$. Three standard mixture pesticide solutions were prepared by diluting each individual standard solution with acetonitrile, in order to get a concentration of about 35 $mg{\cdot}L^{-1}$ for each compound. Working standards solutions were prepared by diluting the three standard mixture pesticide solutions with acetonitrile at a concentration of 1.0 $mg{\cdot}L^{-1}$. All standard solutions were stored in dark below −20°C.

An analytical balance (0.01 mg, Sartorius, Germany), an ultrasonic bath (Type T 760 DH, Prolabo, VWR International, France), a vortex shaker (Model VM-300P, Gemmy industrial corporation, Taiwan), a centrifuge (model U-320R, Boeco, Germany) and a nitrogen sample concentrator (Techne, UK) were used during experiments.

2.2. Solid-Phase Extraction

Two different SPE cartridges were evaluated: Chromabond C_{18}ec, 200 mg, 3 mL, 45 μm from Machery-Nagel (Düren, Germany) and Oasis® HLB (hydrophilic-lipophilic balance) 60 mg, 3 mL, 50 μm from Waters (Milford, MA, USA). The SPE procedure was performed using a VacElut vacuum manifold from Varian (Santa Clara, CA, USA) with a Visiprep large-volume sampler from Supelco (Sigma-Aldrich, St. Louis, MO, USA).

For SPE optimization, blank water samples were spiked with the working standard mixed solution and with the surrogate standard (PCB 52) prepared in MeOH, and then they were shaken manually and with ultrasonic bath to enhance homogenization. The outlines of the SPE procedure used for the extraction of pesticides in water samples were described in our previous paper [10]. The Chromabond C_{18} and Oasis HLB cartridges were preconditioned and activated at the beginning with 3 mL of methanol and then 3 mL of water. Each water sample was percolated using a regulated vacuum through the SPE cartridge with a flow rate of 5 - 10 mL·min^{-1}, and then the cartridge was dried immediately either under vacuum or using centrifugation technique (2500 rpm for 2 min) followed by 10 min under vacuum. The analytes were eluted using the centrifugation technique, where 1 mL of the elution solvent was added to the cartridge with the valve closed, then the cartridge was centrifuged (2300 rpm for 1 min), and finally the elute was recovered (**Figure 1**). This step was repeated if necessary. The extract was collected and evaporated to about 100 µL with a weak nitrogen stream. The obtained extract was reconstituted in 200 µL of acetonitrile, and the solution of recovery internal standards (PCB 153) was added. Samples were stored in the dark at about –20°C until GC-MS analysis.

For method application, groundwater samples were collected in 2.5 L dark glass bottles from three different geographic points located in North Lebanon (Denniyeh), East Bekaa (Terbol) and in South Lebanon (Abbassieh).

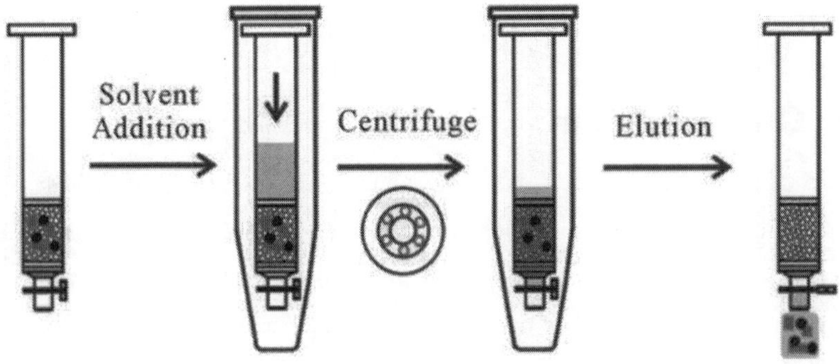

Figure 1. Schematic representation of SPE elution procedure using centrifugation.

Sampling was performed in November 2010 and samples were filtered immediately through a Whatman GF/F filter (0.7 µm porosity), extracted and analyzed.

2.3. GC-MS ANALYSIS

An Agilent 6890N GC connected to an Agilent 5975 MS was used (Agilent technologies, USA). The GC-MS was equipped with Agilent 7683 B autosampler and split/ splitless injector with electronic pressure control. The column used was a capillary column (HP-5MS, 30 m, 0.25 mm i.d., 0.25 μm, Agilent J & W GC columns). The temperature program was the following: initial temperature 70°C, held for 1 min, 10°C·min^{-1} ramp to 160°C then held for 5 min, finally by 3°C·min^{-1} to 240°C and held for 18.5 min. The total analysis time was 60.17 min and the equilibration time 0.5 min. The temperature of the injection port was 250°C and a 1 μL volume was injected in splitless mode. Helium was used as carrier gas at a constant flow of 1 mL·min^{-1}. The mass spectrometer was operated in electron ionization mode with an ionizing energy of 70 eV, ion source temperature 230°C, MS quadruple temperature 150°C, and solvent delay 5.5 min.

Analysis was performed in the selected ion monitoring mode (SIM) based on the use of one target and two or three qualifier ions. Target and qualifier abundances were determined by injection of individual pesticide standards under the same chromatographic conditions in full-scan mode with the mass/charge ratio ranging from m/z 50 to 550. Pesticides were identified according to the retention times, the target and qualifier ions, and the qualifier to target abundance ratios.

2.4. Quantification

When using GC-MS analysis, samples were confirmed to contain a pesticide if the observed retention times of peaks of sample solutions were the same as the peaks for the standards in solvent and the observed abundance ratio of ions was identical to that of the standard ions. Instrumental and procedural blanks were analyzed to avoid laboratory contamination and analytical interferences, and the blank value must be deducted from the results. For the SPE optimization, the quantification was based on peak area ratio of the target ion divided by the internal standard (PCB 153) in order to evaluate the recoveries of the analyzed pesticides. For the method validation, the quantification was done according to the surrogate standard (PCB 52) added from the beginning of the analytical procedure.

3. RESULTS AND DISCUSSION

3.1. Selection of Pesticides

After the study of the pesticides most commonly used in Lebanon, a total of 104 pesticides were initially considered for this study. Under the instrumental conditions given above, efforts were undertaken to optimize the tuning parameters of all the pesticides in the GC-MS in flow injection mode and then their analytical performances were tested in different scan modes. Out of the total pesticides selected, only 67 pesticides were finally found to be suitable for GC-MS analysis. The others were eliminated because either no peaks were observed or their sensitivity was extremely low and did not meet the requirements. Some pesticides were removed due to poor stability; they either decomposed rapidly after preparation of the mixed standard solutions or failed a series of experiments, for example recovery after fortification. Eventually, 67 pesticides from different chemical families including organochlorines, organophosphates, carbamates, pyrethroids, pyrimidines, azoles, triazoles and others, were chosen to be applicable to GC-MS determination. The analyzed pesticides cover a wide range of compounds employed in agriculture, like insecticides, herbicides, fungicides and acaricides. The studied pesticides are listed in **Table 1**.

3.2. Development of the Solid Phase Extraction Procedure

For the optimization of the SPE, several parameters were studied including sorbent selection 1), elution solvent 2), sample volume 3), addition of organic solvent to water sample 4), sorbent drying 5), and elute concentration 6). The complete extraction protocol is outlined in **Figure 2**.

3.2.1. Sorbent Selection (SPE Cartridge)

The pesticide compounds exhibit a wide range of polarities; they fall into the highly polar, medium polar and non-polar category with octanol-water partition coefficient (log K_{ow}) between 0.7 and 7.3 for oxadixyl and bifenthrin respectively. With this in mind, two solid phase sorbents were chosen, Chromabond C_{18} and Oasis HLB. Both sorbents have been applied widely in the research of environmental pollutant chemistry [11,12], and they proved to be suitable for pesticides extraction from water samples [7,13-16].

Preliminary extraction experiments were first conducted with both sorbents to evaluate the efficiency of each cartridge in the analysis of the target pesticides. Three independent recovery assays were performed for each studied cartridge. The elution was done with solvents described in literature: DCM for C_{18} cartridges [17] and MeOH for HLB cartridges [18]. The volume used for elution was (×3) 3 mL. Recoveries showed that both cartridges allow good recovery percentage (higher than 50%) for most of the target compounds, except for tebuconazole, metalaxyl and oxadixyl which showed lower recoveries on C_{18} cartridges (<10%). The poor recovery for the three polar pesticides (log K_{ow} between 0.7 and 3.7) is rather due to the elution step than an adsorption process. Therefore both cartridges, $C_{18}$and HLB, were found suitable for the extraction of the selected pesticides from water and so considered for further optimization in this study.

3.2.2. Solvent Selection for Analyte Elution

The elution strength of the organic solvent depends on the type of sorbent used, so different elution solvents and procedures were evaluated for Chromabond C_{18} and Oasis HLB cartridges. The organic solvents considered in elution procedures were from different polarities including acetone, DCM, Hex, and MeOH.

For traditional elution, a large volume of the elution phase is percolated slowly on SPE cartridge to ensure desorption of analytes from the stationary phase. In our work, the centrifugation technique was applied as an alternate to traditional elution, in order to reduce the volume of the elution phase required. The efficiency of this technique was proved in our previous paper [10], where recoveries obtained using a small volume of elution phase with centrifugation were comparable to those obtained with large volume of elution phase in the traditional elution. The rapid rotation of the centrifuge ensures that the adsorbent is well impregnated with the elution solvent, enabling a complete desorption of the analytes, even the strongly adsorbed, into the solvent. In this new procedure, a small volume (1 mL) of the elution phase was added to the cartridge with the valve closed, then the cartridge was centrifuged (2300 rpm for 1 min), and the elute was recovered. This step was repeated and the total volume of elute was concentrated and analyzed by GC-MS.

For C_{18}, the elution procedures were: (2×) 1 mL of DCM (procedure a), 1 mL of DCM followed by 1 mL of MeOH (procedure b), (2×) 1 mL of DCM/MeOH (1:1, v/v) (procedure c), (2×) 1 mL of Hex/Acetone (1:1, v/v) (procedure d). For HLB, the elution procedures were: (2×) 1 mL of MeOH (procedure e), 1 mL of MeOH followed by 1 mL of DCM (procedure f), (2×) 1 mL of DCM/ MeOH (1:1, v/v) (procedure c), (2×) 1 mL of Hex/Acetone (1:1, v/v) (procedure d). The extracts from each procedure were concentrated, exchanged into 200 μL of acetonitrile and analyzed by GC-MS to determine the recoveries for 50 ng of each pesticide.Results showed that the procedure b (DCM followed by MeOH) provided higher recoveries on average for elution of analytes adsorbed on C_{18}cartridges (92% vs 83% - 84% for procedures a, c, and d; DCM, DCM/MeOH and Hex/Acetone respectively). For the elution of analytes adsorbed on HLB cartridges, similar recoveries of 80% - 83% were obtained for elution using MeOH only and using MeOH followed by DCM (procedures e and f). The procedure f was more preferable as it provides higher recoveries for non-polar analytes, Lambda-cyhalothrin for example. The recoveries for other procedures were lower with averages of 78% for procedure c (DCM/MeOH) and 62% for procedure d (Hex/Acetone).

Table 1. Validation results: linear range, recoveries, repeatability and limits of quantitation (LOQs).

Pesticide	Linearity ($ng \cdot mL^{-1}$)	R^2	C_{18}		HLB		LOQ ($ng \cdot L^{-1}$)
			Recovery (%)	RSD (%)	Recovery (%)	RSD (%)	
Organochlorines							
Aldrin	5 - 500	1.000	89.6	0.6	98.5	1.3	2
p, p'-DDT	10 - 500	0.995	73.3	1.6	75.7	3.1	4
p, p'-DDE	5 - 500	1.000	87.1	2.3	85.3	0.9	2
p, p'-DDD	5 - 500	0.993	87.2	0.2	89.0	2.4	2
Dieldrin	50 - 500	0.999	87.9	1.2	114.3	0.7	20
α-Endosulfan	50 - 500	0.995	95.6	4.6	107.1	1.2	20
β-Endosulfan	50 - 500	0.994	90.9	5.9	101.2	4.6	20
Endosulfan sulfate	50 - 500	0.993	80.5	1.6	83.1	5.2	20
α-HCH	10 - 500	0.999	90.9	2.5	105.9	2.7	4
β-HCH	25 - 500	0.999	92.4	1.3	102.9	1.1	10
γ-HCH (lindane)	25 - 500	0.999	91.0	2.3	114.7	1.4	10
Delta-HCH	50 - 500	0.999	87.4	12.3	89.2	9.7	20
Heptachlor epoxide	25 - 500	1.000	83.5	3.1	112.8	1.1	10
Hexachlorobezene	5 - 500	0.999	84.1	2.0	91.0	2.4	2
Methoxychlor	50 - 500	0.993	91.3	4.0	81.1	2.9	20
Quintozen	25 - 500	0.999	91.6	2.6	92.8	3.0	10

Organophosphates							
Bromophos-ethyl	5 - 500	0.999	84.3	1.6	92.7	1.8	2
Cadusafos	10 - 500	0.999	96.5	2.4	110.7	1.0	4
Chlorpyrifos	25 - 500	0.999	81.7	0.9	100.2	1.7	10
Chlorpyrifos-methyl	5 - 500	0.998	86.9	3.4	96.7	2.3	2
Coumaphos	50 - 500	0.993		11.6	76.4	9.0	20
Diazinon	10 - 500	0.999	90.3	5.1	116.7	2.1	4
Dichlorovos	10 - 500	0.997	103.4	7.9	83.2	5.9	4
Ethion	5 - 500	0.999	103.3	1.6	93.9	1.2	2
Ethoprophos	25 - 500	0.998	96.6	4.2	113.6	1.6	10
Etrimphos	10 - 500	0.998	91.3	3.1	107.1	1.4	4
Fenitrothion	50 - 500	0.995	64.7	1.2	76.8	2.2	20
Malathion	25 - 500	0.995	74.6	6.4	82.0	5.6	10
Methacriphos	10 - 500	0.999	87.6	2.1	115.0	1.1	4
Methidathion	50 - 500	0.995	70.9	7.2	77.6	4.2	20
Parathion	25 - 500	0.996	85.8	5.7	103.9	2.3	10
Parathion-methyl	50 - 500	0.998	82.6	4.6	102.4	3.6	20
Phenthoate	5 - 500	0.998	74.9	1.9	94.5	2.0	2
Phosalone	50 - 500	0.993	83.4	2.5	76.4	6.2	20

Phosphamidon	50 - 500	0.997	83.6	1.3	98.0	2.1	20
Pirimiphos-methyl	25 - 500	0.999	93.6	4.6	109.3	0.9	10
Tolclofos-methyl	5 - 500	0.999	89.7	0.3	105.7	1.2	2
Triazophos	50 - 500	0.996	84.1	5.1	83.4	4.7	20
Carbamates							
Chlorpropham	50 - 500	0.999	81.2	4.6	119.1	2.9	20
Propoxur	5 - 500	0.999	87.8	2.6	114.4	2.0	2
Pyrethroids							
Bifenthrin	5 - 500	1.000	83.2	4.6	82.7	2.5	2
Fenpropathrin	5 - 500	0.999	84.0	6.4	68.0	6.5	2
Lambda-cyhalothrin	50 - 500	0.997	67.5	7.3	68.7	5.8	20
Pyrimidines							
Bupirimate	5 - 500	1.000	93.7	5.6	102.2	3.2	2
Cyprodinil	5 - 500	1.000	103.6	4.5	112.4	1.7	2
Pyrimethanil	5 - 500	1.000	91.9	3.2	117.5	2.1	2
Azoles							
Chlorfenapyr	50 - 500	0.995	86.1	4.2	107.3	3.1	20
Fipronil	25 - 500	0.998	105.8	1.3	114.1	0.2	10
Triazoles							
Myclobutanil	25 - 500	0.999	80.6	1.2	113.2	3.9	10

Penconazole	25 - 500	0.999	90.4	4.6	113.3	3.8	10
Propiconazole	50 - 500	0.999	87.7	1.6	112.7	3.7	20
Tebuconazole	50 - 500	0.998	94.1	4.3	111.2	5.4	20
Other pesticides							
Alachlor	5 - 500	0.999	89.2	2.1	106.3	0.6	2
Boscalid	50 - 500	0.999	84.0	0.6	113.8	2.7	20
Bromopropylate	10 - 500	0.998	82.2	2.4	102.7	3.1	4
Buprofezin	10 - 500	0.999	90.2	1.7	115.5	2.1	4
Fludioxonil	5 - 500	1.000	72.6	2.6	105.4	5.1	2
Kresoxim-methyl	5 - 500	1.000	83.7	1.5	108.2	1.6	2
Metalaxyl	25 - 500	0.999	104.4	3.2	105.8	3.3	10
Oxadixyl	5 - 500	0.998	97.0	4.3	107.7	1.0	2
Pendimethalin	25 - 500	0.998	97.7	4.6	92.1	0.9	10
Procymidone	25 - 500	0.999	92.0	2.0	115.8	2.5	10
Propyzamide	10 - 500	0.999	99.4	4.5	115.6	2.5	4
Tecnazen	10 - 500	0.999	92.7	6.1	110.8	0.6	4
Tetradifon	25 - 500	0.998	85.7	6.4	98.1	1.7	10
Trifluarin	5 - 500	0.997	95.9	2.6	96.0	0.9	2
Vinclozolin	25 - 500	0.999	91.2	3.5	97.9	2.2	10

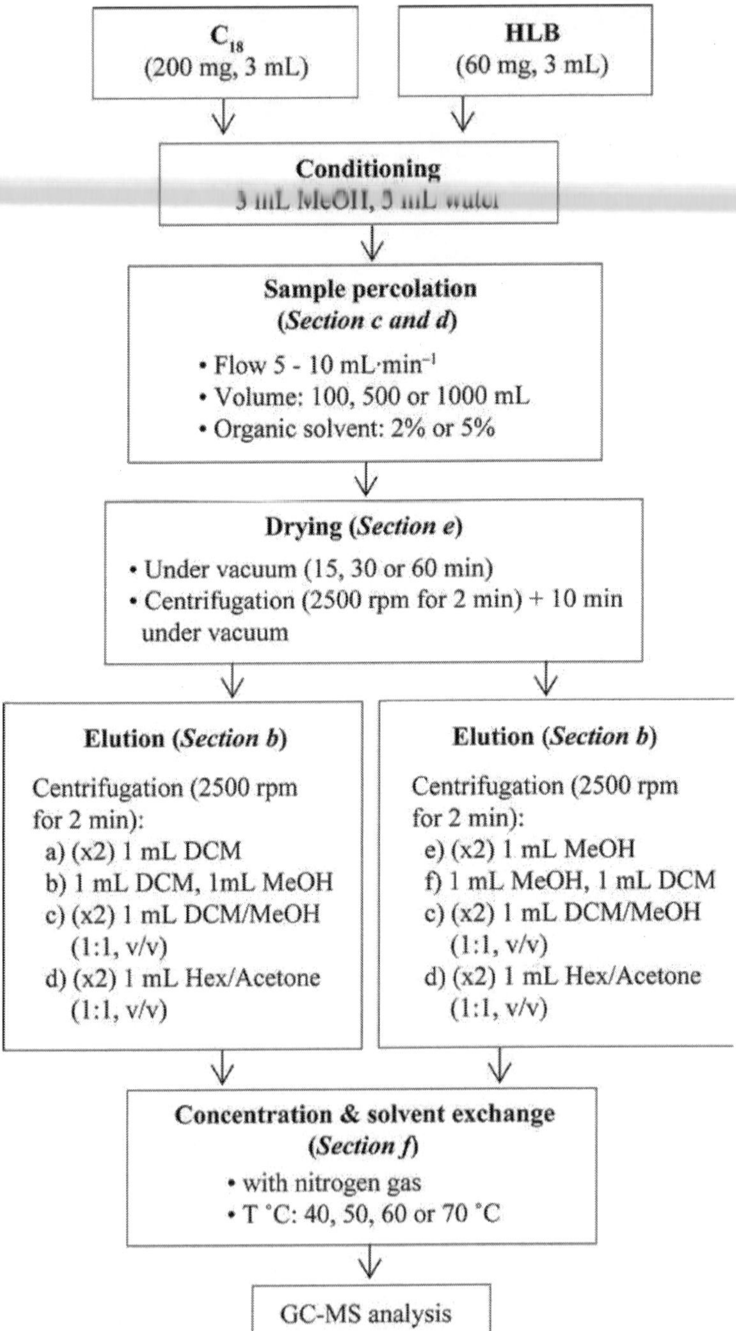

Figure 2. Schematic of the developed analytical method for the water sample extraction and analysis.

3.2.3. Sample Loading Volume

In addition to the selectivity, a higher preconcentration factor is usually expected in the trace level sample analysis. One of the main approaches is changing the volume of water sample to reach the analytical goal. However, large volume of sample will not lead to the maximum extraction factor sometimes because of breakthrough problems, when the analyte is no longer retained by the sorbent due to elution by sample loading volume. Predicting the breakthrough volume based on the analogy between LC and SPE has largely been described [19-21]. The breakthrough volume can vary with the concentration of the target compounds, which is more likely to breakthrough at higher concentration. So the breakthrough volume should be evaluated at the conservatively highest predictive concentration of the target compound in the research area.

In our work, the influence of the sample loading volume was investigated for both cartridges, Chromabond C_{18} and Oasis HLB, where loading volume was varied from 10 to 1000 mL of water fortified with 200 ng·L^{-1}. The results for C_{18} cartridges and HLB cartridges indicated that no apparent differences were found among the enrichment efficiencies for most of the pesticides analyzed over the range of sample volume loaded on C_{18} and HLB cartridges, except for the two highly polar pesticides with high solubility in water, dichlorovos and propoxur (log K_{ow} between 1.54 and 1.9, solubility between 18,000 and 1900 mg·L^{-1} respectively). The recovery of dichlorovos decreased significantly by 34% - 45%, and the recovery of propoxur fell by 4% - 31% for sample volumes of 500 and 1000 mL loaded on C_{18} cartridges. This meant the breakthrough volumes of these two compounds were very low on this type of cartridge. By contrast, the breakthrough volumes of these two compounds and other pesticides in this study exceeded 1000 mL for HLB cartridges based on the increase of recovery with volume observed. Due to comprehensive consideration of enhancing the enrichment factor and avoid excessive sample-handling time and applicability for routine analysis, 500 mL was adopted as the water sample volume for C_{18} and HLB cartridges.

However, low recoveries (lower than 60%) were observed for some target compounds, organochlorines and pyrethroids (p, p'-DDT, p, p'-DDE and p, p'-DDD; bifenthrin, fenpropathrin and lambda-cyhalothrin) for water volumes exceeding 10 mL, which probably resulted from the losses of these hydrophobic compounds induced by the adsorption on the inside wall of the glass sample bottle. This problem is discussed in the next section.

3.2.4. Addition of Organic Solvent

As explained in the last section (5.3.2.c), the addition of organic solvent to water sample was considered in our study to enhance the extraction efficiencies of the high hydrophobic pesticides. Generally, organic solvent addition can increase the solubility of analytes adsorbed on glass bottles in the aqueous phase. MeOH was the organic solvent added to 500 mL of spiked water sample with different percentages (5%, 10% and 20%) and tests were performed on C_{18} and HLB cartridges.

When adding MeOH to water sample, recoveries for most of the analyzed pesticides decreased significantly, especially with a percentage of 10% and 20% of MeOH. This decrease in recoveries was less pronounced for HLB cartridges than for C_{18} cartridges. On the other hand, whatever the percentage of MeOH added, recoveries of the hydrophobic organochlorines and pyrethroids were not strongly improved. Therefore, we decided not to add MeOH to the water samples to prevent losses of analytes due to breakthrough. Low recoveries of the most hydrophobic analytes could be improved by adding the surrogate standard to water sample simultaneously with pesticide standards before SPE extraction. The quantification of pesticides recoveries using the surrogate standard can improve the recoveries of these analytes, and therefore used for the method validation.

3.2.5. Sorbent Drying

Initial sorbent drying after the sample loading was performed on extraction sorbents that had been dried under high vacuum before solvent elution. The influence of the drying time (from 10 to 60 min) was investigated Chromabond C_{18} and Oasis HLB cartridges, after the percolation of 500 mL of blank water fortified with the suite of pesticide compounds (50 ng each). In our previous paper [10], we have shown that the use of the centrifugation technique before vacuum drying ensures the complete remove of residual water from SPE cartridge and reduce the necessary drying time. Therefore, removing water from SPE cartridge using centrifugation technique (2500 rpm for 2 min) followed by a short time vacuum (10 min) was evaluated as well. Results showed that a drying time of 15 or 30 min was not sufficient for C_{18} and HLB cartridges and recoveries of most of analytes improved with drying time to 60 min. Even small amounts of residual water can apparently affect the elution of most of analytes adsorbed on SPE sorbent. When comparing recoveries for cartridges dried for 60 min under vacuum and those dried using centrifugation followed by 10 min under vacuum, we noticed that there was little difference between recoveries. However, it was advantageous to use the centrifugation before vacuum drying in order to reduce the total sample preparation time.

3.2.6. Elute Evaporation Temperature

The effect of elute evaporation temperature was studied to determine the optimum temperature for the organic solvents to vaporize quickly but with smaller amounts of target compounds being lost during the process. Duplicate samples of the elution phase (1 mL DCM + 1 mL MeOH) containing 50 ng of each analyte were concentrated under gentle flow of nitrogen at four temperature levels, including 40°C, 50°C, 60°C and 70°C. For temperatures above 40°C, significant losses occurred for two compounds (fenpropathrin, lambda-cyhalothrin), and there was tend for rest of the analytes among the increasing temperature. Therefore, a temperature of 40°C was used in the final method. For the most volatile analyte, dichlorovos, recoveries werearound 36% for all the evaporation temperature tested.

3.3. Method Validation

The performance of the optimized SPE-GC-MS analysis method was validated evaluating the linearity, recoveries, precision and limits of quantification (LOQs). The results are listed in **Table 1**.

Linearity was determined for the instrumental response. The range of concentration studied was 5 - 500 $\mu g \cdot L^{-1}$ analyzing reference standard solution at seven concentration levels (5, 25, 50, 100, 250 and 500 $\mu g \cdot L^{-1}$).

The calibration curves were plotted using a least-square regression analysis. Each compound showed good linearity for the GC-MS analysis in the studied working range, with correlation coefficient (R^2) greater than 0.993.

The recovery and precision of the developed analytical method were evaluated for C_{18} and HLB cartridges. Recovery studies of all pesticides were performed at 100 ng·L^{-1}, which is the maximal admissible concentration set in the EU directive for individual pesticides in drinking water [4]. Precision was described as the value of relative standard deviation (RSDs) of the areas obtained for each analyte after the replicate (n = 6) analyses of spiked water samples. Recoveries were calculated with the surrogate standard (PCB 52) and values were above 65% - 68% for all the analytes using the C_{18} and HLB cartridges.The overall method repeatability for all pesticides expressed as RSDs was satisfactory and in the range of 0.2% - 12.3% for C_{18} and 0.2% - 9.7% for HLB.

The LOQs were experimentally determined at a signal-to-noise ratio (S/N) of 6, by the spiking of blank water sample at four levels (2, 4, 10 and 20 ng·L^{-1}, n = 6). LOQs obtained were in the low ng·L^{-1} level for all analyzed pesticides (2 - 10 ng·L^{-1}), except for few compounds where a value of 20 ng·L^{-1} was necessary to achieve the LOQ. As a result, this analytical method can be applied in routine for the trace analysis of pesticides in water.

3.4. Application to Real Water Sample

The developed SPE-GC-MS method was applied to the analysis of 67 pesticides in real groundwater samples collected from three different places in Lebanon in November 2010. Water samples were extracted by the developed SPE method according to the optimized conditions described above. For the ground water samples analyzed, no high pesticide concentration was found. Only few compounds were detected at very low concentration in one or more groundwater samples, such as aldrin, DDD, DDE, hexachlorobenzene and diazinon. This means that the presence of these persistent compounds, mainly used in the past, is not critical nowadays and only few traces are still detected in the groundwater system. One pesticide, metalaxyl was detected in the Terbol groundwater sample at a level of 14.5 $ng \cdot L^{-1}$. This compound, mainly used as systemic fungicide to control plant diseases or water-mold fungi, is used on many food and feed crops, and on non-food, residential and greenhouse crops such as tobacco, ornamental plants, trees, shrubs and vines, and lawns and turf. Metalaxyl, moderately stable under normal environmental conditions (with a half-life of 400 days), is persistent and mobile at the same time. His tendency to leach in many soils was demonstrated with a potential to reach groundwater. Metalaxyl has also been detected in ground water in five states in the US at levels typically reaching up to 3 $\mu g \cdot L^{-1}$ [22].

4. CONCLUSIONS

This work presents a new SPE method using the centrifugation technique for the extraction of 67 pesticides from water samples. The proposed method shows practical environmental and economical advantages in terms of sample preparation time, simplicity, reduction in solvent use, and cost and is particularly suitable for routine applications requiring a high sample throughput. Optimized conditions include the percolation of a sample volume of 500 on either C_{18} or HLB cartridges and a drying step using centrifugation followed by 10 min under vacuum. The evaluation of the elution parameters demonstrated that 1 mL of DCM with centrifugation followed by 1 mL of MeOH was successfully used for the elution of analytes on both cartridges. The concentration was performed at a temperature of 40°C. Under the optimized conditions, this method showed good recoveries higher than 65% - 68% for the 67 analyzed pesticides using the C_{18} and HLB cartridges with relative standard deviations lower than 9.7% - 12.3%. Limits of quantification were between 2 and 20 $ng \cdot L^{-1}$. The developed method was successfully applied for the analysis of traces of pesticides in groundwater samples collected from three different regions in Lebanon.

REFERENCES

1 T. Reemtsma and M. Jekel, "Organic Pollutants in the Water Cycle, Wiley-VCH VerlagGmBH& Co. KGaA, Weinheim," Journal of Chromatography A, Vol. 1150, 2006, pp. 267-270.

2 J. A. C. Barth, P. Grathwohl and K. C. Jones, "Introduction to Special Issue Aqua Terra: Pollutant Behavior in the Soil, Sediment, Ground, and Surface Water System," Environmental Pollution, Vol. 148, No. 3, 2007, p. 693.

3 H. Bailey, L. Deanovic, E. Reyes, T. Kimball, K. Larson, K. Cortright, V. Conner and D. E. Hinton, "Diazinon and Chlorpyrifos in Urban Waterways in Northern California, USA," Environmental Toxicology and Chemistry, Vol. 19, No. 1, 2000, p. 82.

4 European Union (EU), "Council Directive 98/83/EC of 3 November 1998 on the Quality of Water Intended for Human Consumption," 1998.

5 European Union (EU), "Directive 2000/60/EC of the European Parliament and of the Council Establishing a Framework for the Community Action in the Field of Water Policy," 2000.

6 J. J. Vreuls, A. J. H. Louter and U. A. Th. Brinkman, "On-Line Combination of Aqueous-Sample Preparation and Capillary Gas Chromatography," Journal of Chromatography A, Vol. 856, 1999, pp. 279-314.

7 M. E. Baez, M. Rodriguez, O. Lastra and P. Contreras, "Solid-Phase Extraction of Organophosphorus, Triazine, and Triazole-Derived Pesticides from Water Samples. A Critical Study," Journal of High Resolution Chromatography, Vol. 20, No. 11, 1997, pp. 591-596.

8 H. Sabik, R. Jeannot and B. Rondeau, "Multiresidue Methods Using Solid-Phase Extraction Techniques for Monitoring Priority Pesticides, Including Triazines and Degradation Products, in Ground and Surface Waters," Journal of Chromatography A, Vol. 885, No. 1-2, 2000, pp. 217-236.

9 M. J. M. Wells and L. Z. Yu, “Solid-Phase Extraction of Acidic Herbicides,” Journal of Chromatography A, Vol. 885, No. 1-2, 2000, pp. 237-250.

10 A. Kouzayha, M. Al Iskandarani, S. Mokh, A. R. Rabaa, H. Budzinski and F. Jaber, “Optimization of a Solid- Phase Extraction Method Using Centrifugation for the Determination of 16 Polycyclic Aromatic Hydrocarbons in Water,” Journal of Agricultural and Food Chemistry, Vol. 59, No. 14, 2011, pp. 7592-7600.

11 M. Asi, A. Hussain and S. Muhmood, “Solid Phase Extraction of Pesticide Residues in Water Samples: Ddt and Its Metabolites,” International Journal of Environmental Research, Vol. 2, No. 1, 2008, pp. 43-48.

12 E. Concha-Grana, M. Turnes-Carou, S. MuniateguiLorenzo, P. Lopez-Mahia, D. Prada-Rodriguez and D. Fernandez-Fernandez, “Evaluation of HCH Isomers And Metabolites in Soils, Leachates, River Water and Sediments of a Highly Contaminated Area,” Chemosphere, Vol. 64, 2006, p. 588.

13 C. De la Colina, A. P. Heras, G. D. Cancela and F. Sanchez Rasero, “Determination of Organophosphorus and Nitrogen-Containing Pesticides in Water by Solid-Phase Extraction with Gas Chromatography with Nitrogen— Phosphorus Detection,” Journal of Chromatography A, Vol. 655, No. 1, 1993, pp. 127-132.

14 United States Environmental Protection Agency (US-EPA), “Method 525: Determination of Organic Compounds in Drinking-Water by Liquid—Solid Extraction and GC-MS”.

15 M. Forcada, J. Beltran, F. J. Lopez and F. Hernandez, “Multiresidue Procedures for Determination of Triazine and Organophosphorus Pesticides in Water by Use of Large-Volume PTV Injection in Gas Chromatography,” Chromatographia, Vol. 51, No. 5-6, 2000, pp. 362-368.

16 Waters Corporation, "Organophosphorus Pesticides in Drinking Water and Fruit. Environmental and Agrochemical Applications Notebook," 2002, p. 9.

17 M. Tahara, R. Kubota, H. Nakazawa, H. Tokunaga and T. Nishimura, "Analysis of Active Oxon Forms of Nine Organophosphorus Pesticides in Water Samples Using Gas Chromatography with Mass Spectrometric Detection," Journal of Health Science, Vol. 52, 2006, pp. 313-319.

18 A. M. Rodrigues, V. Ferreira, V. V. Cardoso, E. Ferreira and M. J. Benoliel, "Determination of Several Pesticides in Water by Solid-Phase Extraction, Liquid Chromatography and Electrospray Tandem Mass Spectrometry," Journal of Chromatography A, Vol. 1150, No. 1-2, 2007, pp. 267-278.

19 M. C. Hennion, C. Cau-Dit-Coumes and V. Pichon, "Trace Analysis of Polar Organic Pollutants in Aqueous samples. Tools for the Rapid Prediction and Optimization of the Solid-Phase Extraction Parameters," Journal of Chromatography. A, Vol. 823, No. 1-2, 1998, pp. 147- 161.

20 M. C. Hennion, "Solid-Phase Extraction: Method Development, Sorbents, and Coupling with Liquid Chromatography," Journal of Chromatography A, Vol. 856, No. 1-2, 1999, pp. 3-54.

21 V. Pichon, "Solid-Phase Extraction for Multiresidue Analysis of Organic Contaminants in Water," Journal of Chromatography A, Vol. 885, No. 1-2, 2000, pp. 195-215.

22 United States Environmental Protection Agency (USEPA), "Metalaxyl. Prevention, Pesticides and Toxic Substances (7508W) EPA-738-F-94-013," 1994.

CHAPTER 3

Product Analysis of Supercritical Fischer-Tropsch Synthesis: Utilizing a Unique On-Line and Off-Line Gas Chromatographs Setup in a Bench-Scale Reactor Unit

Amro Kasht, Rehan Hussain, Minhaj Ghouri, Jan Blank, Nimir O. Elbashir*

Chemical Engineering Program & Petroleum Engineering Program, Texas A & M University at Qatar, Doha, Qatar

ABSTRACT

The utilization of supercritical fluids (SCF) in the Fischer-Tropsch Synthesis (FTS) further complicates the hydrocarbon products identification and analysis process due to the dilution of hydrocarbon peaks by the predominant solvent peak. Therefore, in this project, a custom-made Gas Chromatography (GC) analysis system was designed and implemented to identify and quantify SCF-FTS products. The FTS products were identified using two different methods. The first was through retention time matching by injecting standard solutions, and the second was through the use of the GC/MS system. The quantification of CO and CH_4 was achieved by using external standards, where the CO conversion was calculated by relating the peak area of CO to the peak area of an internal standard (argon) while the CH_4 selectivity was calculated by relating the peak area of CH_4 to that of CO. After setting and calibrating the GC system, two reaction conditions (gas phase: 240°C, 20 bar syngas with 2:1 H_2:CO molar feed ratio and for the supercritical fluids FTS (SCF-FTS): 240°C, 65 bar with 20 bar syngas partial pressure and 2:1 H_2:CO molar feed ratio) were used to compare the different FTS reaction media. The comparison between the gas phase FTS and the SCF-FTS showed the following: carbon monoxide conversion was improved by 14% in the SCF-FTS, while the hydrocarbon product profile SCF-FTS showed 78% reduction in light hydrocarbons (C_1 - C_4) products, 35% increase in middle distillates (C_{11} - C_{22}) products compared to gas phase FTS. These improvements have resulted in higher chain growth probability for the SCF-FTS ($\alpha = 0.85$) compared to the gas phase FTS ($\alpha = 0.76$). These results are generally in agreement with previously reported enhancement in the SCF-FTS [1].

KEYWORDS

Supercritical Fluids, Fischer-Trospch Synthesis, Gas Chromatography, Mass Spectrometry

1. INTRODUCTION

Escalating concerns about the unstable oil prices and oil supply insecurity coupled with stricter environmental regulations have catalyzed the interest in the production of synthetic fuels from alternative resources such as natural gas, coal and biomass via the X-to-Liquid (XTL) process or other process. The Fischer-Tropsch synthesis (FTS) is considered as the heart of the XTL process [2] as it is responsible of converting syngas, a mixture of carbon monoxide (CO) and hydrogen (H_2), into various hydrocarbons that could be refined into environmentally ultra-clean liquid fuels and value added chemicals [3] [4] . The FTS is a surface catalyzed process (heterogeneous catalytic reaction) in which syngas reacts on an active catalyst site; usually cobalt (Co) or iron (Fe) producing longer chain gaseous, liquid and solid (wax) hydrocarbons composed mainly of n-paraffins, α-olefins, isomers, alkenes and oxygenates [5] -[8] .

A forethought towards the desired product selectivity is an extremely important criterion in the design of FTS processes from an economic feasibility point of view. Fixed bed and slurry bubble reactors are the two main reactor types used in industrial Low-Temperature-Fischer-Tropsch (LTFT) [9]. The main drawbacks of fixed bed reactors are the non-uniform temperature distribution along the catalyst bed (axial temperature gradient) and the local hot spots formation that can cause thermal runaway, catalyst deactivation or sintering and rise in CH_4 formation [10] [11] . Slurry reactors were designed to overcome the temperature control issue facing fixed bed reactors, as they operate almost isothermally due to the enhanced heat capacity of the liquid medium (slurry). However, the introduction of the liquid reaction medium induces mass transfer limitations [11]. Therefore, slurry reactors suffer from low conversion rate per pass since the diffusion of syngas in the catalyst pores is slow due to the presence of heavy slurry that fills the catalyst pores [12] [13]. Additionally, catalyst attrition and cumbersome extraction of catalyst from the slurry are also major drawbacks [14]. Consequently, selecting a reactor technology for the FTS is a trade-off since each reactor type and medium has their strengths and weaknesses [15]. Therefore, it makes sense that the idealistic FTS reactor should combine the advantages of commercial FTS reactors while simultaneously eliminates their disadvantages. Such a combination was found possible to some extent through the application of supercritical fluids (SCF) as a FTS reaction media [11] [16]. SCF reaction media was reported to have unique characteristics offering single-phase operation and combined the desired properties of the gas phase (high diffusion) and the liquid phase (high solubility and high heat transfer) [1]. The liquid-like densities and heat transfer properties ensure efficient product desorption and heat removal while gas-like diffusivities ensure high reaction rates and high product removal rate [17].

The analysis and identification of FTS products is quite a cumbersome process since it begins with simple molecules (H_2 and CO) and ends up with hundreds of molecules that can be normal paraffins, isomers, olefins and oxygenates. The introduction of supercritical solvents to the FTS feed stream further complicates the FTS product analysis process due to the dilution of products with the solvent. This dilution results in a large solvent peak that may absorb nearby product peaks and simultaneously result in relatively small hydrocarbons peaks. The groundbreaking study that investigated the application of SCF in the FTS was carried out by Yokota and Fujimoto in 1989 [18]. In their study, they used several on-line Gas Chromatographs (GCs). The first GC had an activated charcoal column and a Thermal Conductivity Detector (TCD) for analyzing CO, CH_4 and CO_2. The second GC utilized a Propak-Q column with a Flame Ionization Detector (FID) to analyze light hydrocarbons. The third GC used two columns (Silicone SE-30 and Dexil-300) with a FID to analyze the hydrocarbons dissolved in the solvent [19]. Dictor and Bell [20] used a single GC with a packed column for the analysis of permanent gases and C_1 - C_5 hydrocarbons. In addition, a capillary column was used for the analysis of C_1 - C_{30+} hydrocarbons. Snavely and Subramaniam [21] also used a single GC similar to that of Dictor and Bell. However, they incorporated a number of modifications from the work of Hackett and Gibbon [22] and Nijs and Jacobs [23] such as the use of a hot trap for the removal of heavy hydrocarbons (wax) before the on-line GC and injecting argon as internal standard for conversion calculations. Furthermore, they have utilized improved capillary and packed columns resulting in better product resolution. Elbashir et al. [24] used two GCs for activity and selectivity calculations. The first GC utilized a capillary column and a FID detector for the analysis of C_2 - C_{40} hydrocarbons while the second GC utilized a packed column and a TCD for analyzing permanent gases (CO, CO_2, H_2, CH_4, C_2H_6, C_2H_4 and N_2). In this project, a custom-made analysis system capable of characterizing gaseous and liquid FTS reactor effluent products was applied. This custom-made GC setup has been designed with the support of Shimadzu. The analysis system provided reliable activity and product distribution data that enabled the validation and study of Supercritical-Fluid-Fischer-Trospch-Synthesis (SCF-FTS). Gaseous FTS products (C_1 - C_6) and permanent gases were analyzed on-line using two GCs in series with TCD and FID while liquid hydrocarbons (C_7 - C_{32})

were analyzed off-line using a Gas Chromatography/Mass Spectroscopy (GC/MS) system.

2. METHODOLOGY

2.1. Catalyst And Reactor Bed

The FTS reported in this work was carried out under both conventional gas phase and non-conventional supercritical and near supercritical phase utilizing the newly commissioned bench-scale novel high-pressure fixed bed reactor unit. The reactions were performed over a cobalt catalyst supported on alumina (15 wt% Co/Al_2O_3) that was prepared using the impregnation technique, and then calcined and reduced under hydrogen flow at elevated temperatures. For the gas phase FTS, the reaction was carried out at 240°C, 20 bar syngas with 2:1 H_2:CO molar feed ratio. As for the SCF-FTS, the reaction was carried out at 240°C, 65 bar with 20 bar syngas partial pressure and 2:1 H_2:CO molar feed ratio. These conditions were selected to provide the basis for comparing the two reaction media. Each experimental set was allowed to reach steady state conditions in terms of activity (CO conversion) and selectivity (hydrocarbon product distribution), as measured by the on-line GC setup. After reaching steady state, liquid samples were collected for off-line analysis.

The cobalt catalyst supported on alumina (15 wt% Co/Al_2O_3) used for the FTS reactions was prepared using the impregnation technique. Firstly, the alumina support (surface area 255 m^2/g, median pore 70 micron and 5000 Å, total pore volume 1.14 cm^3/g, packing density 0.395 kg/L) was crushed, sieved (75 - 150) μm, washed with deionized water and then calcined under airflow at 700°C. Secondly, an aqueous 2 molar cobalt solution was prepared by dissolving cobalt nitrate precursor ($Co(NO_3)_2{\cdot}6H_2O$) in deionized water while stirring at room temperature. Then, the cobalt solution was gradually added to the treated support under constant stirring to avoid lumps formation. Afterwards, the formed catalyst paste was dried, crushed and sieved (150 - 250) μm. Thirdly, the obtained catalyst powder was calcined in a rotary oven under airflow and temperature ramp up to 350°C. Then, the calcined catalyst was diluted with quartz sand by vigorously mixing the catalyst and quartz to assure homogenous catalyst distribution. Finally, different catalyst loadings were added into the reactor tube on top of a layer of glass wool and quartz.

The high-pressure bench-scale reactor unit used in this project consists primarily of five sections (Figure 1). The first section involves the feed delivery setup for gas stream (e.g. syngs, CO, H_2 and argon) and liquid stream (e.g. solvent hexane). The second section is the reaction zone where the FTS reaction takes place. The third section utilizes two separation columns (hot trap and cold trap) to separate permanent gases, liquid hydrocarbons and wax. The fourth section is composed of the reactants and products on-line GC analysis system. The fifth section, which is the utility section, is composed of three sub-sections: 1) pneumatic air, 2) chilled water system and 3) ventilation setup.

At the feed delivery section syngas (33.4% CO and 66.6 H_2), carbon monoxide (99.97% CO), hydrogen (99.999% H_2) and argon (99.999% Ar) were introduced into the system at controlled flow rates (0 - 500 nmL/min) using four mass flow controllers (MFCs). For the supercritical phase testing, liquid solvent, n-hexane (98.5% n-hexane) was introduced at controlled volumetric flow rates using a High Pressure Liquid Chromatography (HPLC) pump. A feed purification system composed of a cylindrical vessel filled with three layers of adsorbers/catalysts (BASF Selexsorb COS, BASF PuriStar R3-15 and BASF E-315) and an inline gas filter were employed at the solvent stream and at each gas stream for the removal of COS, H_2S, CS_2, CO_2, acetylene, O_2, arsine, sulfur compounds, moisture and particulates. The liquid solvent is then passed through a vaporizer before it is mixed with the feed gases in the static mixer that ensure proper mixing of the feed gases and the SCF.

The reaction mixture then enters the custom made tubular fixed bed reactor (Autoclave Engineers, 16" overall length, 12" heated length, 2/3" internal diameter, 1.0" outer diameter, 73 cm^3 net volume in heated zone). The reactor tube was vertically embedded in a hollow ceramic insulated electric furnace with three heating zones that allow reactor temperature control within ±2.0°C. The pressure control valve located between the reactor and the hot trap controlled the reactor pressure.

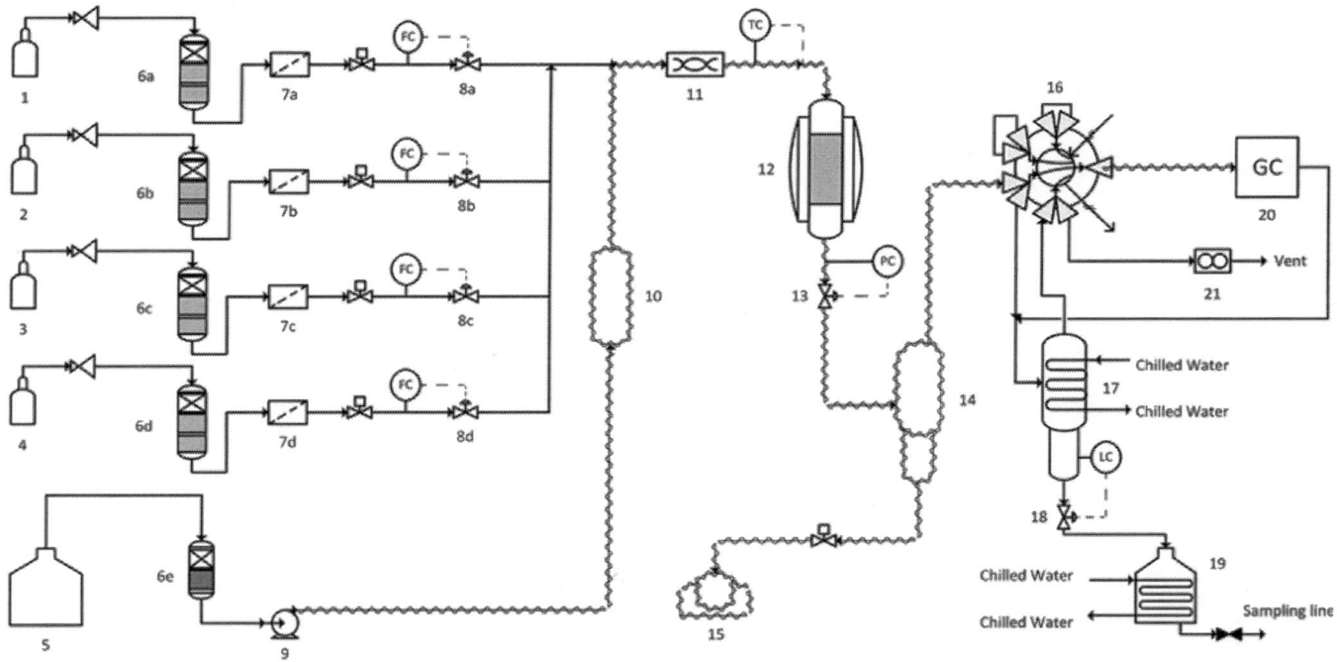

Figure 1. Process Flow Diagram (PFD) of the high-pressure reactor unit. (1) Syngas cylinder, (2) Hydrogen cylinder, (3) Carbon monoxide cylinder, (4) Argon cylinder, (5) Solvent storage tank, (6 a-e) Feed purification system, (7 a-d) Inline filter, (8 a-d) Mass flow controllers, (9) Solvent pump, (10) Vaporizer, (11) Static mixer, (12) Fixed bed reactor, (13) Pressure control valve, (14) Hot trap, (15) Wax collection vessel, (16) GC injection valve, (17) Cold trap, (18) Level control valve, (19) Product collection tank, (20) On-line GC analysis system, (21) Flow meter.

Two flash separation columns were placed downstream of the high-pressure reactor. The first one is the hot trap surrounded by a heating tape and insulated by glass wool and aluminum foil. The hot trap is operated at 150°C and 0.9 barg to facilitate the condensation of heavy hydrocarbons (wax) prior to the on-line GC system. The formed wax is collected in a special pressure vessel for future analysis. All the lines connecting the reactor with the hot trap, the wax collection vessel and the on-line GC were heated to 150°C to prevent any wax condensation.

The remaining hot trap gas stream goes towards the cold trap used to separate liquid hydrocarbons and permanent gases. The cold trap is internally cooled using cooling coils with chilled water flow at 4.0°C. Hydrocarbons and water condensed by the cold trap were collected in a tank and liquid samples were periodically collected for off-line GC analysis.

In a typical experimental run, the system was first flushed with n-hexane and then purged with nitrogen. Secondly, the catalyst was activated in-situ under hydrogen flow (100 - 350 nmL/min) at a temperature range between 180°C to 350°C. Thirdly, the system was purged and flushed again. Fourthly, the solvent was introduced (2.43 - 2.74 nmL/min) and the reactor pressure (20 - 80 bar) and temperature (230°C - 240°C) were slowly ramped to their desired set points. After the temperature and pressure had been stabilized syngas was allowed (50 - 150 nmL/min) at the desired flow rate. Each experimental set was allowed to reach steady state conditions in terms of conversion and product distribution, as indicated by the on-line GC. After reaching steady state, wax and liquid samples were collected for further off-line analysis.

2.2. Gas Chromatography Setup

A fraction of the gases leaving from the top of the hot trap is directed towards the on-line GC system through an air actuated 8 way selecting valve that is used to inject samples at specified intervals. Operation of the sampling valves was controlled by a custom-made Shimadzu GC Postrun software that acquires samples at preset times and record TCD and FID data. This system has been designed to ensure that the same sample to be injected simultaneously to two combined GCs with three detectors. This unique on-line analytical system consists of dual gas chromatographs with three detectors, TCD-1, FID and TCD-2 (Figure 2). TCD-1 detector analyzes permanent gases (O_2, N_2, CH_4, CO, CO_2, C_2 and H_2S) by utilizing two GC columns. Both columns use helium as a carrier gas at 5.0 mL/min and 4.0 bar. The FID detector analyzes (C_3 - C_{13}) hydrocarbons by utilizing a single fused silica capillary column to separate the hydrocarbons. The capillary column also uses helium as a carrier gas at 5.0 mL/min and 4.0 bar, while air (350 mL/min, 4.0 bar) and H_2 (35 mL/min, 4.0 bar) were used to light the detector flame and N_2 as makeup gas at 25 mL/min and 4.0 bar. Hydrocarbon products heavier than C_{15} were not detected by the on-line FID. One more detector TCD-2 is used with a single packed column to detect H_2, where in this column N_2 was used as a carrier gas at 25 mL/min and 4.0 bar.

Liquid products from the cold trap are periodically collected for off-line analysis using GC/MS system equipped with FID and Mass Spectroscopy (MS) that uses a capillary column to separate (C_5- C_{30}) hydrocarbons. The column uses helium as carrier gas at 5.0 nmL/min and 4.0 bar, air at 350 nmL/min and 4.0 bar, and H_2 at 35 nmL/min and 4.0 bar to light the FID flame and N_2 as makeup gas at 25 nmL/min and 4.0 bar.

2.3. Activity, Selectivity and Product Distribution Calculations

The CO inlet flow rate to the FTS reactor was measured and controlled by the MFC; however, CO participates in the FTS reaction, and thus its outlet flow rate cannot be known directly. In order to be able to calculate the outlet CO flow rate, it was necessary to use an inert (argon) as an internal standard. The method adopted herein is similar to that previously used by Nijs and Jacobs [23] . Argon does not participate in the FTS reaction, and thus its inlet and outlet flow rates are equal under steady-state conditions. Hence, it can be used to calibrate the outlet CO flow rate by relating the CO peak area/Ar peak area ratio (found using on-line GC TCD channel 2) to CO flow rate. The trend line (straight-line equation) of the peak area ratio

(PA_{CO}/PA_{Ar}) against the CO flow rate can then be used to calculate the CO outlet flow rate as follows:

$$Q_{CO,out} = \frac{\left[\left(\frac{PA_{CO}}{PA_{Ar}}\right)+b\right]}{a} \quad (1)$$

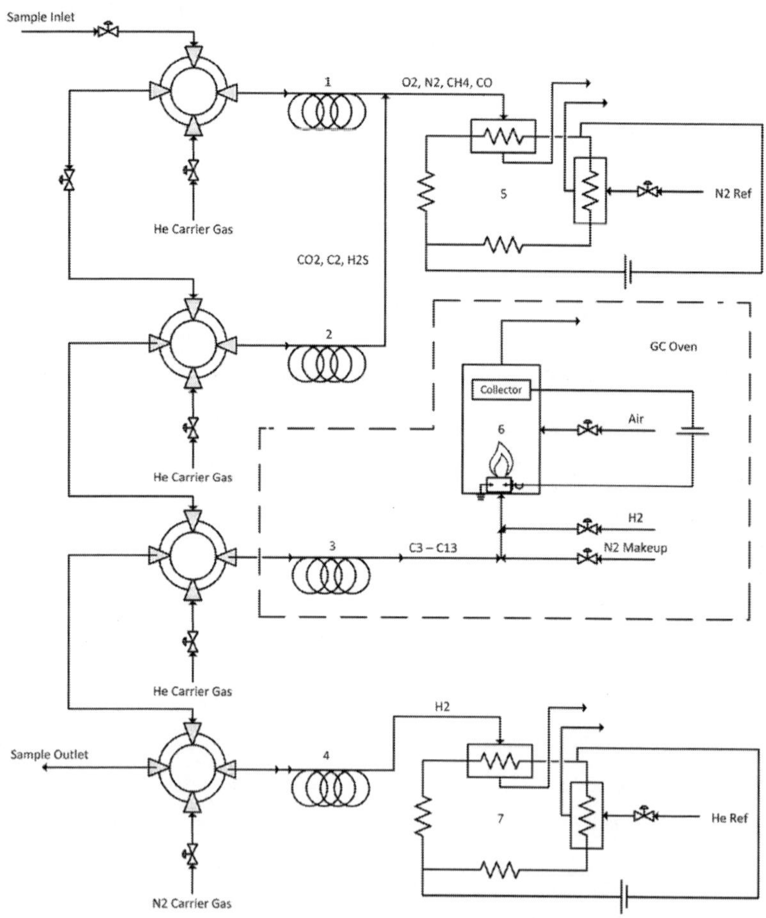

Figure 2. Process Flow Diagram (PFD) of the custom made Shimadzu on-line GC analysis system. (1) Restek Molesieve 13X packed column, (2) Restek Molesieve 5A packed column, (3) Restek Rtx-1 capillary column, (4) Restek Porapak Q packed column, (5) TCD-1, (6) FID, (7) TCD-2.

where $Q_{CO,out}$ is the outlet flow rate of CO, "a" is the slope and "b" is the intercept. For example, syngas with H_2:CO of 2:1 (CO is 1/3 of syngas flow) was allowed into the system at different flow rates (10 - 200) nmL/min. Argon was allowed at constant flow rate (10 or 15) nmL/min. The resulting gas mixture was then directed toward the on-line GC/TCD, where CO and Ar peak areas were determined. CO and Ar peak ratio (PA_{CO}/PA_{Ar}) was then calculated and plotted against CO flow rate. Figure 3 shows the calibration line for argon flow rate of 15 nmL/min. The resulting straight line equation (y = 0.0548x – 0.0264), where y = PA_{CO}/PA_{Ar} and x = $Q_{CO,Out}$ can be rearranged to find the outlet CO flow rate as follows:

$$Q_{CO,out} = \frac{\left[\left(\frac{PA_{CO}}{PA_{Ar}}\right) + 0.0264\right]}{0.0548} \qquad (2)$$

The material balance calculation was based on carbon, where the amount of carbon (as CO) entering the reactor is equal to the amount of carbon (as unreacted CO and produced hydrocarbons) leaving it. Therefore, the CO consumption rate and conversion throughout the experimental run were calculated based on the difference between moles of CO entering the reactor and moles of unreacted CO leaving the reactor as shown in Equation (3) and Equation (4). Note that syngas H_2:CO ratio of 2:1 was normally used unless CO, H_2 were introduced separately.

$$x_{CO} = \left(\frac{Q_{syngas,in}}{3} + Q_{CO,in}\right) - Q_{CO,out} \qquad (3)$$

$$X_{CO} = \frac{x_{CO}}{\left(\frac{Q_{syngas,in}}{3} + Q_{CO,in}\right)} \times 100\% \qquad (4)$$

where x_{CO} is the consumption rate of CO and X_{CO} is the conversion percent of CO.

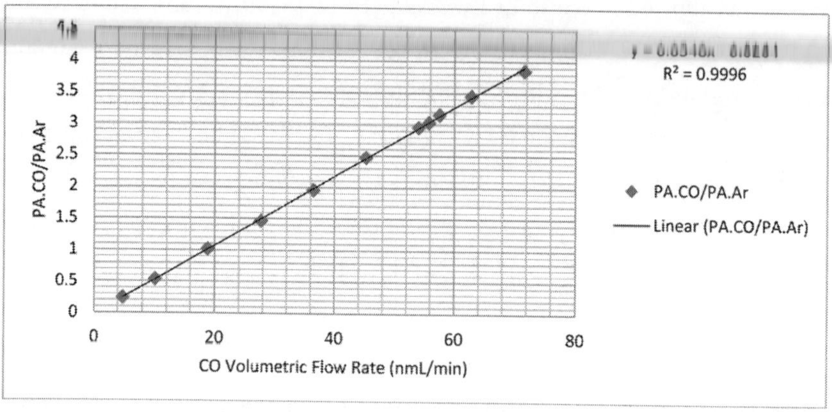

Figure 3. CO/Ar calibration curve at 15 nmL/min Ar flow rate, used to calculate outlet CO flow rate.

In order to calculate the CH_4 outlet flow rate, a calibration gas with 10.09 CO mol·% and 4.00 CH_4mol·% (n_{CH_4}/n_{CO} molar ratio = 0.397) was used to find the responding peak area for each component and relate CO flow rate to CH_4 flow rate. The peak areas for CO and CH_4 for the aforementioned concentrations were identified by the on-line GC-TCD and the CH_4/CO peak area ratio $\left(PA_{CH_4}/PA_{CO}\right)$ is measured as 0.331. After that, the molar ratio and peak area ratio were used to calculate CH_4 outlet flow (CH_4 formation rate) as follows:

$$Q_{CH_4,out} = \left(\frac{PA_{CH_4,out}}{PA_{CO,out}}\right) \cdot \left[\frac{\left(\frac{n_{CH_4}}{n_{CO}}\right)}{\left(\frac{PA_{CH_4}}{PA_{CO}}\right)}\right] \cdot Q_{CO,out} \tag{5}$$

CH_4 selectivity was then calculated as follows:

$$S_{CH_4} = \frac{Q_{CH_4,out}}{\left(\frac{Q_{syngas,in}}{3} + Q_{CO,in} - Q_{CO,out}\right)} \times 100\% \tag{6}$$

where S_{CH_4} is the selectivity towards methane.

The hydrocarbon product distribution was determined utilizing the on-line and off-line FID GC setups. The hydrocarbon formation rate ($g/g_{cat}{\cdot}h$) was calculated as the sum of formation rate of all species (isomers, normal paraffins, olefins, alcohols) with "n" carbon number. Then, the normalized weight percentages (W_n) of each carbon number was calculated from the ratio of the hydrocarbon formation rate of carbon number "n" to the total weight of hydrocarbons formed as detected by the FID.

It was difficult to determine the exact amount of C_6 hydrocarbons produced when running the FTS reaction under near and supercritical conditions since large quantities of n-hexane were present in the feed. This resulted in a very large n-hexane peak absorbing nearby peaks within the C_6 range. Thus, calculating the formation rate of C_6 hydrocarbons was impractical. Hence, the hydrocarbons within the C_6 product range were eliminated from the product distribution and were not included in the Anderson-Schulz-Flory (ASF) calculations. Additionally, for consistency purposes, C_6hydrocarbons were also eliminated in the gas phase FTS when comparing the performance under the two reaction mediums.

To start the hydrocarbon product distribution calculations, a reference or starting point was needed. For the on-line GC, this was the CH_4 formation rate ($g/g_{cat}{\cdot}h$), previously determined using the GC TCD. The TCD CH_4 formation rate with the CH_4 peak area given by the FID was used as a reference ratio and held constant for each sample. The ratio was then used to calculate the formation rate of hydrocarbons with different carbon numbers. This method is similar to that previously used by Snavely and Subramaniam [21] . As for the off-line GC FID analysis, a similar approach was used, however the CH_4 formation rate was replaced with n-C_7 formation rate

($g/g_{cat}\cdot h$), determined using the on-line FID, and the on-line n-C_7 formation rate with the off-line n-C_7 peak area was used as a reference ratio in this case.

As discussed below in section 3.1, each hydrocarbon present in the FID spectrum was identified and assigned with its distinct carbon number and type (isomer, normal, olefin, alcohol). After that, the peak area for each carbon number with the same type was summed, and then the total peak area for each carbon number was summed. The weighted molecular weight for each carbon number (n) was then estimated as follows:

$$\text{Weighted } MW = \frac{\sum\left(PA_{n,\text{Type}} \cdot MW_{n,\text{Type}}\right)}{\text{Total } PA_n} \tag{7}$$

The total hydrocarbon formation rate for each carbon number (n), was then calculated as follows:

$$F_{\text{HC}} = \frac{\left[\left(\dfrac{Q_{\text{CH}_4,\text{out-TCD}}}{PA_{\text{CH}_4\text{-FID}}}\right) \cdot \text{Total } PA_n\right]}{\left[\dfrac{\sum\left(PA_{n,\text{Type}} \cdot MW_{n,\text{Type}}\right)}{\text{Total } PA_n}\right]} \tag{8}$$

where F_{HC} is the hydrocarbon formation rate. $Q_{\text{CH}_4,\text{out-TCD}}$ is the methane formation rate estimated by the on- line TCD, $PA_{\text{CH}_4\text{-FID}}$ is the peak are of methane given by the on-line FID.

The weight fraction of hydrocarbons containing "n" carbon atoms (W_n) was then calculated as follows:

$$W_n = \frac{F_{HC,n}}{\sum[F_{HC}]} \quad (9)$$

where $F_{HC,n}$ is the hydrocarbon formation rate for "n" carbon number.

The natural logarithm of the weight fraction of carbon number "n" over the carbon number $[\ln(W_n/n)]$ was calculated in order to plot the ASF distribution, where $[\ln(W_n/n)]$ was plotted against "n". The slope of the trend line can then be used to find the chain growth probability as follows:

$$\ln\left(\frac{W_n}{n}\right) = [\ln(\alpha)] \cdot n + \ln\left[\frac{(1-\alpha)^2}{\alpha}\right] \quad (10)$$

where ln(α) is the slope of the trend line and the chain growth probability is equal to the exponential of the slope ($\alpha = e^{slope}$).

3. RESULTS AND DISCUSSION

3.1. Hydrocarbons Product Identification

The exact and detailed identification of FTS products can be hindered by the large number of compounds involved [21] . One method for identifying FTS products is to use standard solutions that contain FTS products (n-paraffins, α-olefins, β-olefins, branched and oxygenates of C_1 - C_{34}) to obtain a satisfactory GC calibration. This is a far too complex and cumbersome process; thus a GC/MS analyzer was used for compound identification in the range of C_5 - C_{34}.

The FTS products under both conventional gas phase and supercritical phase were found to be qualitatively similar (but not quantitatively, as will be discussed later). The main difference between the GC product distributions under the two reaction mediums is the dilution of the FTS products in the SCF-FTS, as illustrated and compared in Figure 4 and Figure 5. This dilution complicates the identification process since hydrocarbons have a smaller relative peak area and are more difficult to detect when the solvent/syngas ratio is increased. Thus, gas phase FTS samples were primarily used for the product identification and then the product identification was validated for the SCF-FTS samples and found to be in good agreement.

For the identification of C_1 - C_4 hydrocarbons and permanent gases (CO, CO_2, H_2, Ar) a number of standard gas mixtures with known molar percentages were used. The standard gas mixtures included common FTS products within the (C_1 - C_4) range. These hydrocarbons include methane, ethylene, ethane, propylene, propane, 1-butene, n-butane, iso-butane, butylene, butane, as well as H_2 and CO_2. In addition, different solutions containing known concentrations of (C_1 - C_6) normal alcohols (methanol, ethanol, propanol, butanol, pentanol and hexanol) were injected into the on-line GC for identifying the alcohol peaks. After that, a calibration table was prepared by entering the retention time for each identified species into the GC data collection software. The identification of permanent gases is shown in Figure 6 and sample peak identification for C_4 hydrocarbon group with species fingerprints is shown in Figure 7. As illustrated in Figure 6, for the TCD 1 signal (channel 2), helium was used as the carrier gas. Thus, any gas slower than C_2 was back-flushed. The peak order obtained from the detector as shown in Figure 6 started with H_2 (small peak), Argon and O_2(they do not separate), N_2, CH_4, CO, ethylene, ethane and CO_2.

The identification of C_5-C_{34} hydrocarbon compounds were carried out off-line by injecting liquid FTS product samples into the GC/MS (Figure 8). The off-line GC/MS analysis of liquid products provided the full product distribution upon accurate identification of all hydrocarbons between C_5 and C_{34}. Most hydrocarbons (isomers, olefins and oxygenates) between C_5 and C_{25} were easily identified with high accuracy by the GC/MS. However, higher molecular weight branched isomers and high molecular weight olefins were more problematic to identify. At that point the remaining peaks were identified via superposition since FTS products were observed to follow a similar pattern for each hydrocarbon group. Huff et al. [25] used a similar GC/MS and superposition technique for identifying FTS products. However, they limited their GC/MS product identification and verification to major hydrocarbons and oxygenates within C_1 - C_{12} hydrocarbons. After that, they used superposition for identifying the remaining peaks.

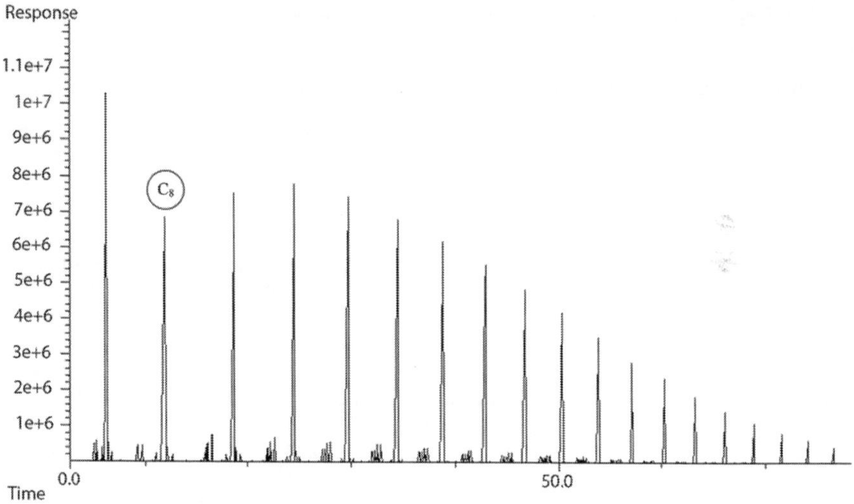

Figure 4. Overall FTS product spectrum obtained using off-line FID analysis, for cobalt-based, gas phase FTS at 240°C and 20 bar with H_2:CO of 2:1 feed ratio.

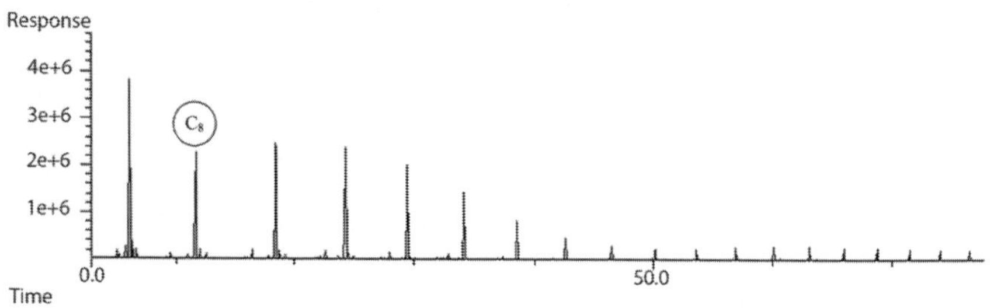

Figure 5. Overall FTS product spectrum obtained using off-line FID analysis, for cobalt-based, supercritical phase FTS at 240°C and 65 bar with H2:CO of 2:1 feed ratio.

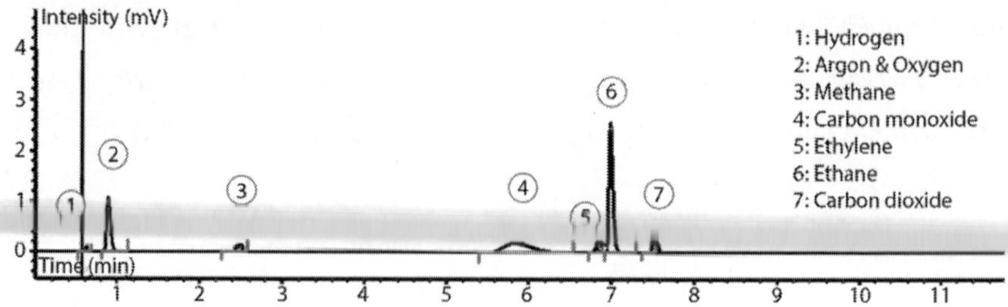

Figure 6. Peak identification of permanent gases on the TCD.

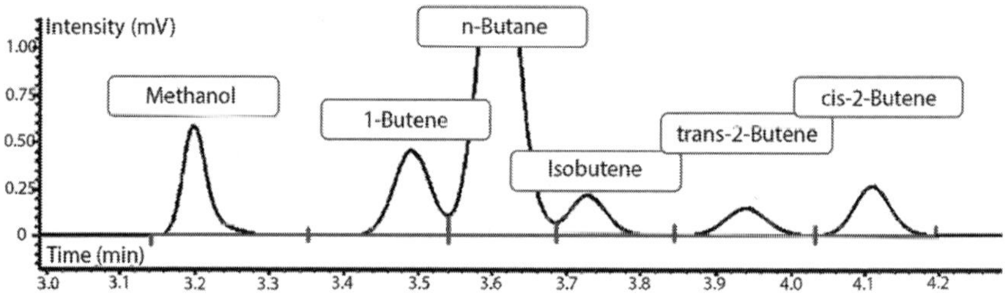

Figure 7. Peak identification for C_4 FTS product range.

For the product distribution and selectivity studies, FTS products were divided into five groups (alcohols, isomers, α-olefins, normal paraffins and alkenes). As previously mentioned, the FTS products followed a similar pattern for each hydrocarbon group (C_X), where it begins with (C_X) isomers (increasing in number as X increases). The peaks follow the aforementioned C_X of α-olefins (from 1 - 3), the normal (C_X) alkane (normally the peak with the highest area, which is common for cobalt-based catalyst). After the normal alkane peak comes, the cis and then the trans (C_X) alkenes peaks, as demonstrated in Figure 9 and Figure 10. On the other hand, the location of (C_{X-3}) alcohol peaks changed as the carbon number increased. For lighter alcohols (C_3 - C_7), the alcohol peak was located before the C_X isomers, for (C_8 - C_{13}) alcohols, the alcohol peaks were located between C_X isomers and for (C_{14} - C_{20}) alcohols, the alcohol peaks were found after the isomers and before the α-olefins peaks. It is also noticeable that, α-olefins were not detectable by the Agilent GC/MS beyond (C_{17}), alcohols beyond 1-eicosanol (C_{20} normal alcohol) were not detectable, while cis and trans alkenes were not detectable beyond (C_{22}). All observations mentioned earlier agree well with the profile of FTS hydrocarbon product distribution obtained from cobalt-based catalysts tested under similar conditions [21] [24]. A sample of compounds (C_7 range) identified using the GC/MS is shown in Table 1. The sample retention time on both FID and MS were identified while the molecule identity was determined utilizing the MS (database), as shown in Figure 11. Subsequently, the retention time for each compound was inserted manually into the GC software (Enhanced Data Analysis), for automatic report generation with defined components.

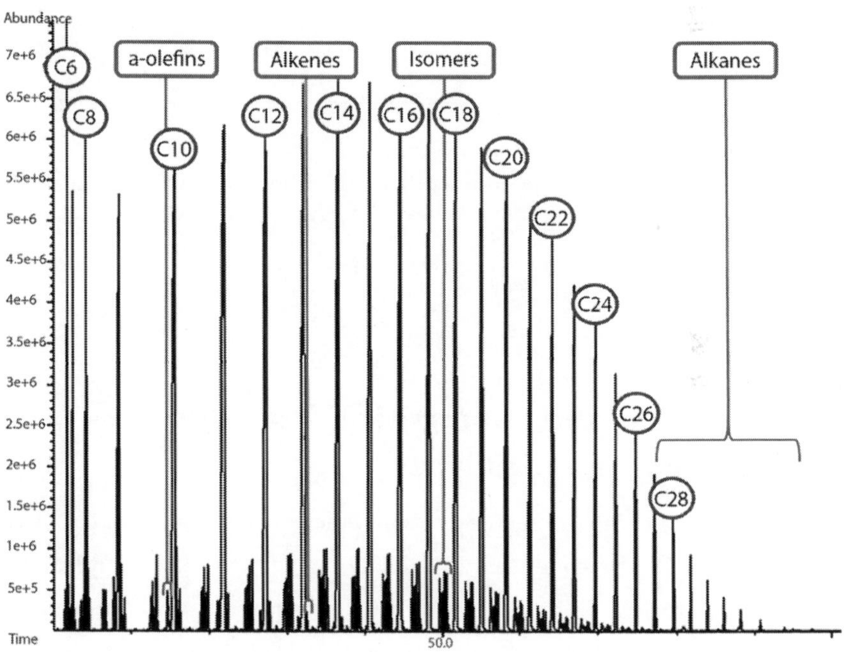

Figure 8. Overall FTS product spectrum obtained using off-line MS analysis, for cobalt-based, gas phase FTS at 240°C and 20 bar with H_2:CO of 2:1 feed ratio.

During the off-line GC/MS FTS product identification minor aromatic, cyclic and acidic compounds were identified in the organic phase. This is expected since these compounds have higher affinity and would be more dissolved in the aqueous phase. Since these compounds were not prevalent in the FTS product and the error resulting by ignoring them would be minimal, they were not included in the FTS product distribution. For instance, the combined total weight percent of aromatics, acids and esters in the organic FTS phase was estimated by Hackett et al. [26] to be less than 4 wt%. While, acids and esters constituted about 8 wt% of the aqueous FTS samples as estimated by Anderson and White [27].

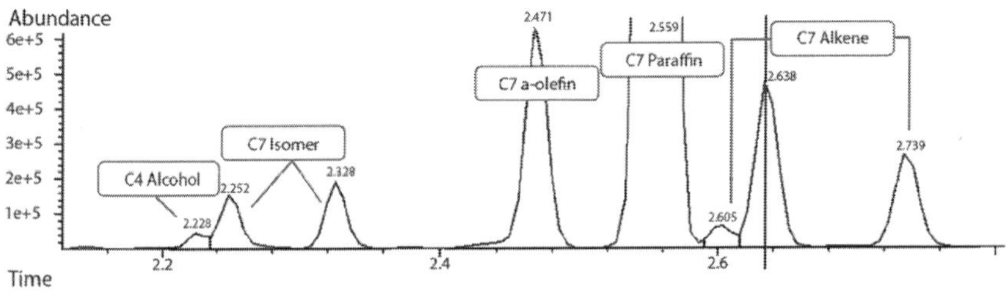

Figure 9. FTS product range on the MS, as identified by the off-line GC/MS, for cobalt-based, gas phase FTS at 240°C and 20 bar with H_2:CO of 2:1 feed ratio.

3.2. Gas Phase and Supercritical Phase Fischer-Tropsch Synthesis Study

In this section, a comparison study between the conventional gas phase FTS and the SCF-FTS was conducted to validate the FTS product identification and analysis process. This was carried by investigating the enhancements in terms of activity, selectivity and chain growth probability (α-value). The reactor was operated at 20 bar and 240°C for the gas phase FTS. On the other hand, for the SCF-FTS the supercritical solvent (n-hexane) was introduced at 240°C and the total reactor pressure was maintained at 65 bar. In both cases, the syngas partial pressure was kept at 20 bar with syngas with H_2:CO of 2:1 feed ratio.

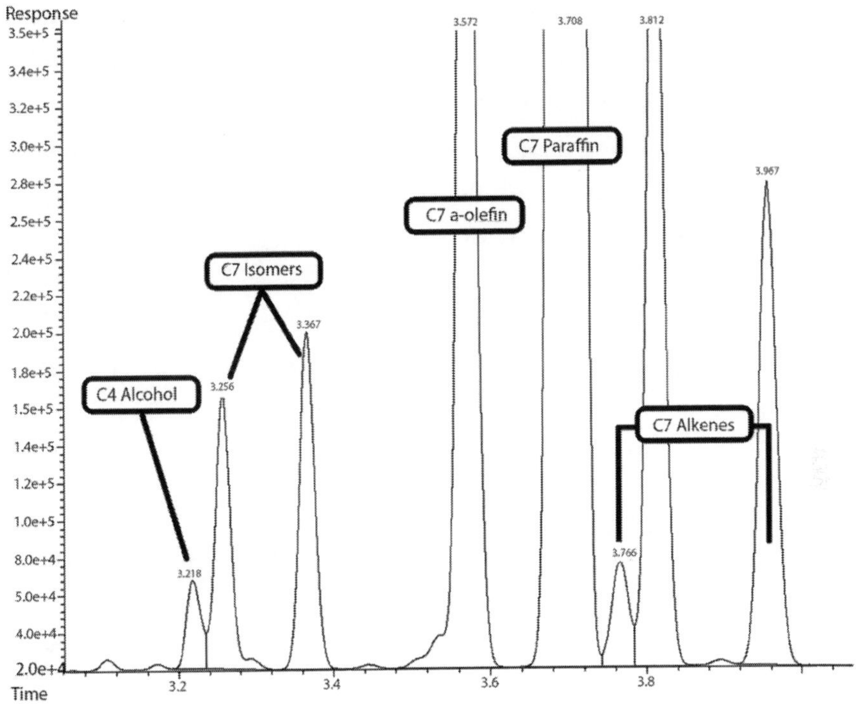

Figure 10. C_7 FTS product range on the FID, as identified by the off-line GC/MS, for cobalt-based, gas phase FTS at 240°C and 20 bar with H_2:CO of 2:1 feed ratio.

The on-line and off-line GC FTS product analysis confirmed a yield of a complex mixture of C_1 to C_{34} hydrocarbons and C_1 to C_{17} oxygenates, in addition to H_2O, CO_2 and unreacted syngas. The predominant FTS products were linear alkanes (n-paraffins), while branched (isomers), oxygenates (alcohols), α-olefins, and alkenes (cis and trans) hydrocarbons were produced to a lesser extent, as previously shown in Figures 8-10; this product distribution is typical for cobalt-based catalysts.

Starting with Figure 12, which compares the CO conversion percent under the two reaction media, it was noticed that CO conversion increased by approximately 14% upon the introduction of supercritical hexane. In literature [18] [28] [29] , the experimental data concerning CO conversion under the different reaction media is contradictory. On one hand, Fujimoto et al. [18] found that the conversion under supercritical condition was to some extent lower than that in the gas phase. They attributed this to the difference in syngas diffusion rate in the different reaction media where the diffusion rate in supercritical fluids is slower than that in the gas phase. On the other hand, Huang and Roberts [30] showed that conversion in supercritical hexane is higher (70%) than that in the gas phase (60%).

Table 1. C_7 FTS product range as identified by the off-line GC/MS, for cobalt-based, gas phase FTS at 240°C and 20 bar with H_2:CO of 2:1 feed ratio.

FID Retention Time (min)	Peak Area (a.u.)	MS Retention Time (min)	Compound Name	Compound Group	Compound Branch
3.218	612,711	2.228	1-butanol	C_4	Alcohol
3.258	2,065,458	2.252	2-methyl hexane	C_7	Isomer
3.367	2,459,938	2.328	3-methyl hexane	C_7	Isomer
3.572	10,665,767	2.471	1-heptene	C_7	α-Olefin
3.708	154,553,134	2.559	n-heptane	C_7	Normal
3.766	855,344	2.605	3-methyl-2-hexane-Cis	C_7	Alkene
3.812	7,000,586	2.638	2-heptene-Cis	C_7	Alkene
3.957	4,094,594	2.739	2-heptene-Trans	C_7	Alkene

The overall hydrocarbon product distributions from C_1 - C_{32} under the previously mentioned conditions were compared as shown in Figure 13. The overall product distribution figure elucidates the shift towards the production of heavier hydrocarbons under the SCF reaction medium. For instance, middle distillates (C_{11} - C_{22}) constituted about 67% of the normalized product weight under the SCF-FTS compared to 44% in the gas phase FTS. In addition, a substantial drop in methane selectivity was observed when shifting the FTS reaction medium from gas phase to SCF. A number of studies [18] [28] [31] [32] also reported a suppression of methane formation when applying SCF as a reaction media for the FTS.

Figure 11. Example of MS peak identification for C_7 peak # 4 (1-heptene) using the off-line GC/MS, for cobalt-based, gas phase FTS at 240°C and 20 bar with H_2/CO feed ratio of 2.

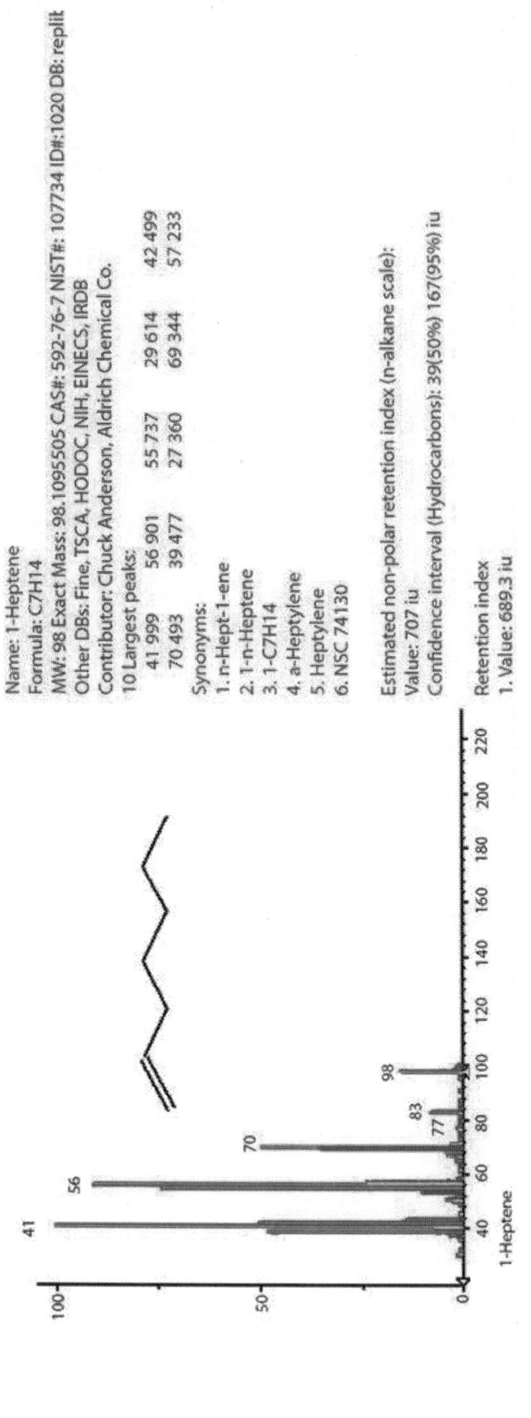

Higher α-value indicates higher selectivity towards heavier hydrocarbon production and vice versa.Figure 14 shows the ASF plots of the FTS under either reaction media (supercritical hexane and gas phase). From the ASF plots, it was first noticed that lower selectivities towards methane and light hydrocarbons were achieved under supercritical conditions, while the selectivity towards middle distillates and heavy hydrocarbons was increased (this is in agreement with the previous discussion). The improved product selectivities upon the introduction of supercritical hexane resulted in a significant increase in chain growth probability from 0.76 to 0.85. Similar incensement were observed in previous literature [17] [24] [33] [34] under supercritical conditions.

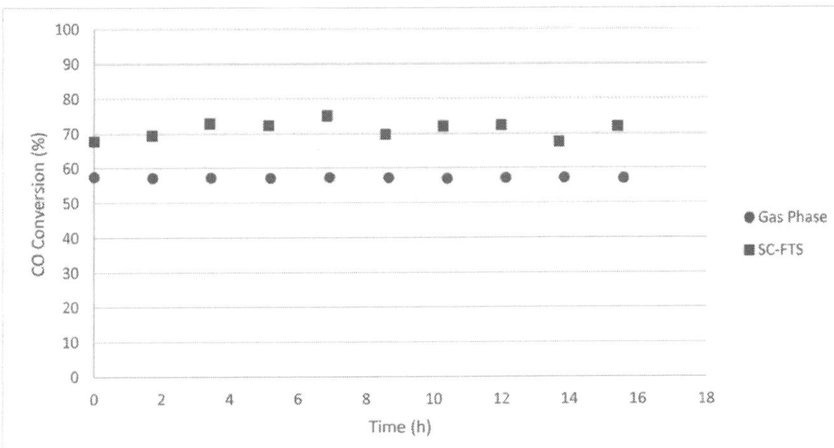

Figure 12. Reaction medium effect on CO conversion, over 15 wt% Co/Al_2O_3, under the following conditions: temperature 240°C, pressure 65 bar for SC-FTS and 20 bar for gas phase FTS, and H_2:CO of 2:1 feed ratio.

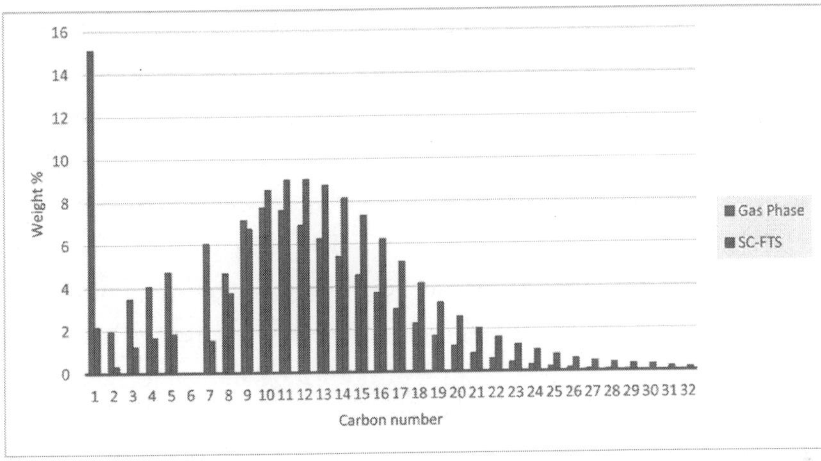

Figure 13. Reaction medium effect on product distribution, over 15 wt% Co/Al_2O_3, under the following conditions: temperature 240˚C, pressure 65 bar for SC-FTS and 20 bar for gas phase FTS, and H_2:CO of 2:1 feed ratio.

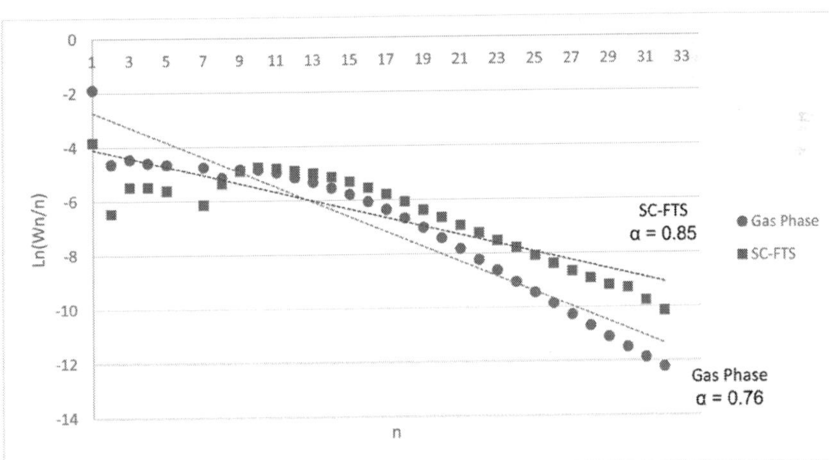

Figure 14. Anderson?Schulz?Flory (ASF) plots showing the reaction medium effect on the hydrocarbon product distribution and chain growth probability (α-value), over 15 wt% Co/Al_2O_3, under the following conditions: temperature 240˚C, pressure 65 bar for SC-FTS and 20 bar for gas phase FTS, and H_2:CO of 2:1 feed ratio.

The aforementioned enhancements in CO conversion, reduced selectivity towards light hydrocarbons and increased selectivity towards heavy hydrocarbons is explained in literature and is discussed briefly in this paper. The increased conversion is attributed to the increased reactants accessibility to the active sites after the in-situ wax extraction from the catalyst pores by the SCF [1] [16] [17] [28] [31] [35] -[37] . The reduction in methane and light end fuel cuts is attributed to the liquid-like heat transfer properties in the SCF reaction medium that results in better heat removal from the catalyst bed. Thus, it minimizes hot spots formation and reduces the rate of cracking reactions [16] [28] [31] [32] [38] . As for the observed enhancement in chain growth probability and the selectivity towards heavy hydrocarbons, this is explained by the liquid-like solubility of SCF that result in higher diffusivities and desorption rates of heavy hydrocarbons (wax extraction) from the catalyst pores. Thus, freeing more actives sites for primary products to incorporate in the chain growth process [1] [16] [17] [28] [31] [35] [37] . Moreover, SCF are capable of extracting olefins prior to undergoing hydrogenation reactions (termination of the chain growth), which increase the chance of their readsorption and incorporation in the chain growth, thus yielding heavier FTS products [1] [16] [17] [31] [33] -[35] .

3.3. Alcohol Analysis Verification

During the off-line FTS product peak identification using the (GC/FID/MS) system, considerable amounts of several alcohols were identified in the organic FTS product samples. This is uncommon for cobalt-based low-temperature FTS; thus, further investigation was required.

A typical liquid sample of the gas phase FTS (240˚C, 20 bar, 65 nmL/min syngas flow with 2:1 H_2:CO molar feed ratio) was collected from the bottom of the cold trap. Then the organic phase from that sample was injected into the GC/MS for analysis. The alcohols identified within the sample's organic phase are given in Table 2.

To further investigate these results, the aqueous phase from the same sample was injected into the GC/MS, where the resulting MS spectrum is shown in Figure 15, labelled with the identities of the peaks, as determined by the

MS. The GC/MS analysis of the aqueous phase clearly shows that C_1 - C_6 alcohols were present in the FTS product.

As a double check and to confirm the MS results, a reference sample was prepared by mixing a small amount (~5 μL) of a mixture of C_1 - C_6 alcohols diluted with 1.5 mL of n-hexane (Sigma Aldrich, ≥95% purity, with the rest composed of small amounts of C_6 iso and cyclic hydrocarbons). The reference sample was then injected into the GC/MS, and the obtained MS spectrum is shown inFigure 16. By comparing Figure 15 and Figure 16, it can be concluded that the reference sample confirms that the GC/MS correctly identified the alcohol peaks.

The FTS produces a wide range of products including n-paraffins, olefins oxygenates and aromatics. Pei et al. [39] recently studied the promotional effect of silica over activated carbon supported cobalt catalyst for the production of higher chain alcohols. They reported that sufficient amounts of SiO_2 addition favored the formation of higher molecular mass alcohols (C_6 - C_{18}) by suppressing cobalt reduction and therefore producing Co(II) species and thus facilitating CO insertion. The same group previously also demonstrated that activated carbon supported cobalt catalysts are able to produce linear C_1 - C_{18} alcohols from syngas via the FTS reaction due to the synergy between active Co_2C sites formed in-situ during the reaction and active cobalt sites [40] . Also, another experimental work [41] reported CO conversion of 30% and high alcohol selectivity of 20% when cobalt molybdenum based catalyst is used under the operating conditions of 307°C, 75 bar and syngas ratio $H_2/CO = 2$.

From the above discussion, it can be seen that while iron based catalyst are known to be selective towards alcohol production, cobalt based catalyst can also produce alcohols if the catalyst and reactions conditions were adjusted. Nevertheless, this finding requires further detailed investigation, which is currently being performed by the research team.

Table 2. Peak list for the alcohols identified in the gas phase FTS (240°C, 20 bar, 65 nmL/min syngas flow with H_2:CO of 2:1 molar feed ratio) organic phase liquid sample, injected into GC/FID/MS system.

Alcohol	Retention time (min)	Response (a.u.)
1-Propanol	2.457	276,269
1-Butanol	3.203	467,519
1-Hexanol	9.666	4,132,448
1-Heptanol	16.473	25,329,552
1-Octanol	22.644	36,927,076
1-Nonanol	28.049	41,910,385
1-Decanol	32.923	33,100,899
1-Undecanol	37.417	32,615,101
1-Dodecanol	41.619	31,647,273
1-Tridecanol	45.578	29,650,492
1-Tetradecanol	49.331	27,495,369
1-Pentadecanol	52.892	22,176,038
1-Hexadecanol	56.279	13,716,154

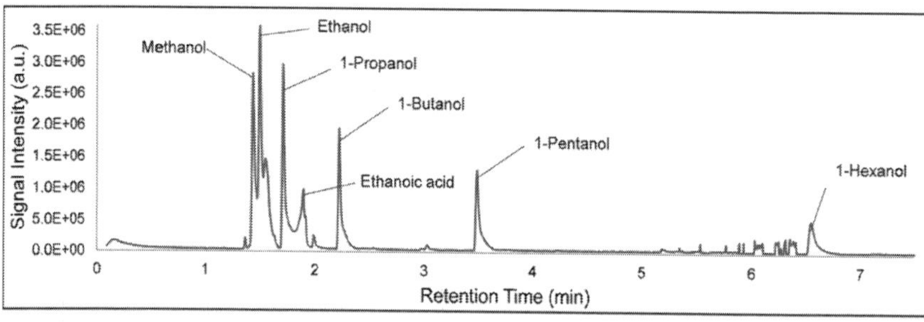

Figure 15. GC/MS spectrum for the aqueous FTS (240°C, 20 bar, 65 nmL/min syngas flow with 2:1 H^2:CO molar feed ratio) sample, with compound identities labeled.

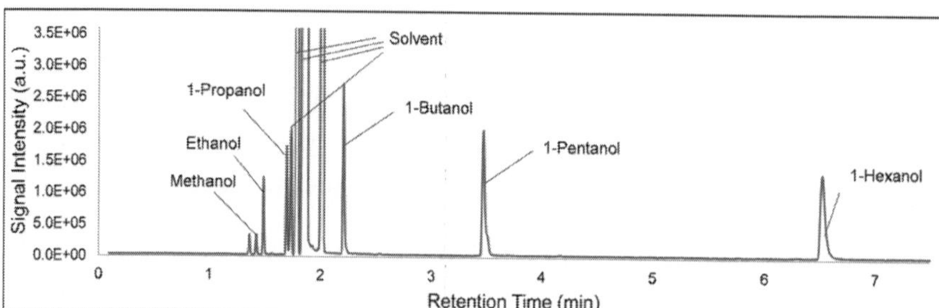

Figure 16. GC/MS spectrum of the prepared reference sample, consisting of C_1 - C_6 alcohols dissolved in n-hexane, with compound identities labelled.

4. CONCLUSION

The dilution of FTS product with the solvent complicated the identification and analysis of the reactor effluent since the solvent peak was extremely dominant over the product spectrum. Therefore, a sophisticated custom-made on-line and off-line GC setup was developed. The GC analysis system and method implemented successfully enabled the identification and characterization of FTS hydrocarbons within (C_1 - C_{32}) range as well as permanent gases (CO, H_2, CO_2, Ar). After setting the analysis system, a set of reaction conditions were applied to verify the functionality of the GC analysis setup. Through the utilization of the custom-made analysis system, reliable activity and product distribution were obtained, which enabled the formulation of conversion, product distribution and ASF curves under both gas phase and supercritical phase. These curves were able to compare the cobalt catalyst performance under the distinct reaction media. The experimental findings provided us with accurate measurements of the hydrocarbon product distribution both the conventional gas-phase and in the supercritical phase, which are in good agreement with the product profiles reported in literature. This unique analytical setup has as well provided us the opportunity to identify and confirm the presence of significant alcohol amounts in the product profile of a cobalt-based catalyst, which are not common products for this type of catalyst. This phenomenon is currently under thorough investigation in our lab to understand the mechanistic of reaction that could lead to such products.

ACKNOWLEDGEMENTS

The authors would like to thank the Qatar National Research Fund (QNRF) (a member of Qatar Foundation) for financial support of this project under the National Priorities Research Program (NPRP) (grant # 4-144-2-590). The statements made herein are solely the responsibility of the authors.

CITE THIS PAPER

AmroKasht,RehanHussain,MinhajGhouri,JanBlank,Nimir O.Elbashir, (2015) Product Analysis of Supercritical Fischer-Tropsch Synthesis: Utilizing a Unique On Line and Off-Line Gas Chromatographs Setup in a Bench-Scale Reactor Unit. *American Journal of Analytical Chemistry*,**06**,659-676.

REFERENCES

1. Elbashir, N.O., Bukur, D.B., Durham, E. and Roberts, C.B. (2009) Advancement of Fischer-Tropsch Synthesis via Utilization of Supercritical Fluid Reaction Media. AIChE Journal, 56, 997-1015.

2. Fischer, F. and Tropsch, H. (1926) The Synthesis of Petroleum at Atmospheric Pressures from Gasification Products of Coal. Brennstoff-Chemie, 7, 97-104.

3. Marvast, M.A., Sohrabi, M., Zarrinpashne, S. and Baghmisheh, G. (2005) Fischer-Tropsch Synthesis: Modeling and Performance Study for Fe-HZSM5 Bifunctional Catalyst. Chemical Engineering Technology, 28, 78-86.

4. Sheehan, J., Camobreco, V., Duffield, J., Graboski, M. and Shapouri, H. (1998) Life Cycle Inventory of Biodiesel and Petroleum Diesel for Use in an Urban Bus. U.S. Department of Energy, U.S. Department of Agriculture, Colorado.

5. Elbashir, N.O. and Eljack, F.T. (2010) A Method to Design an Advanced Gas-to-Liquid Technology Reactor for Fischer-Tropsch Synthesis. Proceedings of the 2nd Annual Gas Processing Symposium, Doha, 369-377.

6. Steynberg, A. and Dry, M. (2004) Fischer-Tropsch Technology: Studies in Surface Science and Catalysis, Vol. 152. Elsevier, New York.

7. Hussain, R., Blank, J.H., Todic, B., Elbashir, N.O. and Bukur, D.B. (2014) Development of Gas-to-Liquid Technologies from Micro- to

Mac-ro-Scale. In: Excellence and Impact of Research, Texas A&M University at Qatar, Doha, 55-78.

8. Moutsoglou, A. and Sunkara, P.P. (2011) Fischer-Tropsch Synthesis in a Fixed Bed Reactor. Energy & Fuels, 25, 2242-2257.

9. Jager, B. (2003) Development of Fischer Tropsch Reactors. The AIChE Spring Meeting, New Orleans, 1 April 2003, 1-9.

10. Lee, T.S. and Chung, J.N. (2012) Mathematical Modeling and Numerical Simulation of a Fischer-Tropsch Packed Bed Reactor and its Thermal Management for Liquid Hydrocarbon Fuel Production using Biomass Syngas. Energy & Fuels, 26, 1363-1379.

11. Elmalik E., Tora, E., El-Halwagi, M.M. and Elbashir, N.O. (2011) Solvent Selection for Commercial Supercritical Fischer-Tropsch Synthesis Process. Fuel Processing Technology, 92, 1525-1530.

12. Maretto, C. and Krishna, R. (2001) Design and Optimisation of a Multi-Stage Bubble Column Slurry Reactor for Fischer-Tropsch Synthesis. Catalysis Today, 66, 241-248.

13. Stern, D., Bell, A.T. and Heinemann, H. (1983) Effects of Mass Transfer on the Performance of Slurry Reactors Used for Fischer-Tropsch Synthesis. Chemical Engineering Science, 38, 597-605.

14. Hooshyar, N., Vervloet, D., Kapteijn, F., Hamersma, P.J., Mudde, R.F. and v. Ommen, J.R. (2012) Intensifying the Fischer-Tropsch Synthesis by Reactor Structuring—A Model Study. Chemical Engineering Journal, 207-208, 865-870.

15. Bartholomew, C.H. and Farrauto, R.J. (2006) Fundamentals of Industrial Catalytic Processes. 2nd Edition, Wiley, Hoboken.

16. Yokota, K., Hanakata, Y. and Fujimoto, K. (1990) Supercritical Phase Fischer-Tropsch Synthesis. Chemical Engineering Science, 45, 2743-2750.

17. Bochniak, D.J. and Subramaniam, B. (1998) Fischer-Tropsch Synthesis in Near-Critical n-Hexane: Pressure Tuning Effects. AIChE, 44, 1889-1896.

18. Yokota, K. and Fujimoto, K. (1989) Supercritical Phase Fischer-Tropsch Synthesis Reaction. Fuel, 68, 255-256.

19. Yokota, K. and Fujimoto, K. (1991) Supercritical-Phase Fischer-Tropsch Synthesis Reaction. 2. The Effective Diffusion of Re-actant and Products in the Supercritical-Phase Reaction. Industrial & Engineering Chemistry Research, 30, 95-100.

20. Dictor, R.A. and Bell, A.T. (1984) On-Line Analysis of Fischer-Tropsch Synthesis Products. Industrial & Engineering Chemical Fundamentals, 23, 252-256.

21. Snavely, K. and Subramaniam, B. (1997) On-Line Gas Chromatographic Analysis of Fischer-Tropsch Synthesis Products Formed in a Supercritical Reaction Medium. Industrial & Engineering Chemical Research, 36, 4413-4420.

22. Hackett, J.P. and Gibbon, G.A. (1989) Capillary Gas Chromatographic Characterization of Fischer-Tropsch Liquefaction Products: On-Line Analysis. Journal of Chromatographic Science, 27, 405-412.

23. Nijs, H.H. and Jacobs, P.A. (1980) On-Line Single Run Analysis of Effluents from a Fischer-Tropsch Reactor. Chromatographic Science, 19, 40-45.

24. Pei, Y., Ding, Y., Zhu, H. and Dua, H. (2015) One-Step Production of C1 - C18 Alcohols via Fischer-Tropsch Reaction over Activated Carbon-Supported Cobalt Catalysts: Promotional Effect of Modification by SiO2. Chinese Journal of Catalysis, 36, 355-361.

25. Pei, Y., Ding, Y., Zang, J., Song, X., Dong, W., Zhu, H., et al. (2014) Fischer-Tropsch Synthesis: Characterizing and Reaction Testing of Co2C/SiO2 and Co2C/Al2O3 Catalysts. Chinese Journal of Catalysis, 36, 252-259.

26. Zhao, Y. (2007) On the Investigation of Alcohol Synthesis via the Fischer Tropsch Reaction. Doctor of Philosophy, Chemistry, Cardiff University, United Kingdom.

27. Fan, L. and Fujimoto, K. (1999) Fischer-Tropsch Synthesis in Supercritical Fluid: Characteristics and Application. Applied Catalysis A: General, 186, 343-354.

28. Elbashir, N.O. (2010) Opportunities for Selective Control of Fischer-Tropsch Synthesis Hydrocarbons Product Distribution. 2nd Annual Gas Processing Symposium, Doha, 10-14 January 2010, 197-208.

29. Irankhah, A. and Haghtalab, A. (2008) Fischer-Tropsch Synthesis Over Co-Ru/γ-Al2O3 Catalyst in Supercritical Media. Chemical Engineering Technology, 31, 525-536.

30. Subramaniam, B. (2001) Enhancing the Stability of Porous Catalysts with Supercritical Reaction Media. Applied Catalysis A: General, 212, 199-213.

31. Elbashir, N.O., Dutta, P., Manivannan, A., Seehra, M.S. and Roberts, C.B. (2005) Impact of Cobalt-Based Catalyst Characteristics on the Performance of Conventional Gas-Phase and Supercritical-Phase Fischer-Tropsch Synthesis. Applied Catalysis A: General, 285, 169-180.

32. Tsubaki, N., Yoshii, K. and Fujimoto, K. (2002) Anti-ASF Distribution of Fischer-Tropsch Hydrocarbons in Supercritical-Phase Reactions. Journal of Catalysis, 207, 371-375.

33. Yan, S., Fan, L., Zhang, Z., Zhou, J. and Fujimoto, K. (1998) Supercritical-Phase Process for Selective Synthesis of Heavy Hydrocarbons from Syngas on Cobalt Catalysts. Applied Catalysis A: General, 171, 247-254.

34. Geerlings, J.J.C., Wilson, J.H., Kramer, G.J., Kuipers, H.P.C.E., Hoek, A. and Huisman, H.M. (1999) Fischer-Tropsch Technology—From Active Site to Commercial Process. Applied Catalysis A: General, 186, 27-40.

35. Huang, X. and Roberts, C.B. (2003) Selective Fischer-Tropsch Synthesis over an Al2O3 Supported Cobalt Catalyst in Supercritical Hexane. Fuel Processing Technology, 83, 81-99.

36. Bukur, D.B., Lang, X. and Nowicki, L. (2005) Comparative Study of an Iron Fischer-Tropsch Catalyst Performance in Stirred Tank Slurry and Fixed-Bed Reactors. Industrial & Engineering Chemical Research, 44, 6038-6044.

37. de-Deugd, R.M., Kapteijn, F. and Moulijn, J.A. (2003) Trends in Fischer-Tropsch Reactor Technology—Opportunities for Structured Reactors. Topics in Catalysis, 26, 29-39.

38. Anderson, R.R. and White, C.M. (1994) Analysis of Fischer-Tropsch By-Product Waters by Gas Chromatography. Separation Science, 17, 245-250.

39. Hackett, J.P., Gibbon, G.A. and Feldman, J.A. (1985) Capillary Gas Chromatographic Characterization of Fischer-Tropsch Liquefaction Product Oils. Chromatographic Science, 23, 285-292.

40. Huff, G.A., Satterfield, C.N. and Wolf, M.H. (1983) Stirred Autoclave Apparatus for Study of the Fischer-Tropsch Synthesis in a Slurry Bed: 2 Analytical Procedures. Industrial and Engineering Chemistry Fundamentals, 22, 258-263.

41. Elbashir, N.O. and Roberts, C.B. (2005) Enhanced Incorporation of α-Olefins in the Fischer-Tropsch Synthesis Chain-Growth Process over an Alumina-Supported Cobalt Catalyst in Near-Critical and Supercritical Hexane Media. Industrial & Engineering Chemistry Research, 44, 505-521.

CHAPTER 4

Resolution of Volatile Fuel Compound Profiles from *Ascocoryne Sarcoides*: A Comparison by Proton Transfer Reaction-Mass Spectrometry and Solid Phase Microextraction Gas Chromatography-Mass Spectrometry

Natasha D Mallette,[1,2] W Berk Knighton,[1] Gary A Strobel,[1] Ross P Carlson [1,2] and Brent M Peyton[1,2]

[1]*Department of Chemical and Biological Engineering, Montana State University*

[2]*Center for Biofilm Engineering, Montana State University*

ABSTRACT

Volatile hydrocarbon production by *Ascocoryne sacroides* was studied over its growth cycle. Gas-phase compounds were measured continuously with a proton transfer reaction-mass spectrometry (PTR-MS) and at distinct time points with gas chromatography-mass spectrometry (GC-MS) using head space solid phase microextraction (SPME). The PTR-MS ion signal permitted temporal resolution of the volatile production while the SPME results revealed distinct compound identities. The quantitative PTR-MS results showed the volatile production was dominated by ethanol and acetaldehyde, while the concentration of the remainder of volatiles consistently reached 2,000 ppbv. The measurement of alcohols from the fungal culture by the two techniques correlated well. Notable compounds of fuel interest included nonanal, 1-octen-3-ol, 1-butanol, 3-methyl- and benzaldehyde. Abiotic comparison of the two techniques demonstrated SPME fiber bias toward higher molecular weight compounds, making quantitative efforts with SPME impractical. Together, PTR-MS and SPME GC-MS were shown as valuable tools for characterizing volatile fuel compound production from microbiological sources.

KEYWORDS

Biofuel Solid phase microextraction Proton transfer reaction-mass spectrometry Volatile organic compounds Fungal hydrocarbons Gas chromatography-mass spectrometry

1. INTRODUCTION

Ascocoryne sarcoides is an endophytic fungus recently isolated from Northern Patagonia and has the potential to produce a petroleum-like fuel (Mycodiesel®) directly from a cellulose fermentation process [Strobel, Knighton, Kluck, Ren, Livinghouse, Griffin, Spakowicz and Sears 2008]. Fungi produce medium to long chain hydrocarbons, and the prevalence of these compounds is in the C_{19}-C_{30} chain length [Ladygina, Dedyukhina and Vainshtein 2006]. In contrast, *Ascocoryne sarcoides* (NRRL 50072) has been reported to produce a series of straight chained and branched medium chain-length hydrocarbons of C_5-C_{10} chain length, in the range of gasoline fuel, including heptane, 2-pentene, octane, 1-methyl-cyclohexene, 3,5-octadiene, and cyclodecene [Solomons and Fryhle 2002, Griffin, Spakowicz, Gianoulis and Strobel 2010]. This interesting metabolism requires further work to characterize growth patterns and develop the organism for potential biofuel applications.

Recent studies have identified a variety of compounds produced by *A. sarcoides* on different media using headspace GC-MS on discrete solid phase microextraction (SPME) fiber samples [Griffin, Spakowicz, Gianoulis and Strobel 2010, Strobel et al. 2008]. The SPME technique was introduced in the 1990's [Arthur and Pawliszyn 1990], and SPME fibers have been used to identify and quantify complex volatile mixtures including hydrocarbons from water, soil, food and wine [Langenfeld et al. 1996, Mallouchos et al. 2002, Parkerton et al. 2000, Robinson et al. 2011]. Its advantages include minimizing systematic errors with extraction of volatiles, eliminating the need for solvents [Wercinski 1999], and being simple, fast and relatively sensitive [Pawliszyn 1999]. However, several studies have revealed instabilities with quantification by this method [Brás et al. 2011, Weldegergis et al. 2011, Zeng and Noblet 2002], so further evaluation of the quantitative use of the method for volatile mixtures from *A. sarcoides* was tested.

Another analytical method with potential for analyzing biofuels, proton transfer reaction-mass spectrometry (PTR-MS) has been used to monitor volatile organic compounds (VOCs) emitted from plants, bacteria and subterranean fungi [Aprea et al. 2007, 2009, Bäck et al. 2010,Bunge et al. 2008, Kai et al. 2010, Maleknia et al. 2007, Ramirez et al. 2009, Steeghs 2004]. These studies often employed solid substrata, rarely included submerged liquid fungal cultivation [Bäck et al. 2010], and collected data in discrete time points. To our knowledge, however, no studies have used PTR-MS to investigate the continuous evolution of volatile organic fuel compounds from a submerged liquid fungal culture over its entire growth cycle.

An advantage of PTR-MS is that it can be run in real time to continuously collect data on the concentrations of specific compounds [Ezra et al. 2004, Lindinger et al. 1998, Tomsheck et al. 2010]. PTR-MS is primarily a quantitative tool and generally requires a detailed knowledge of the components present to be successfully employed. Ion mass (as measured by a quadrupole mass spectrometer) is not a unique indicator of chemical composition. For instance, an ion at m/z 107 can represent either $C_8H_{10}H^+$ (ethylbenzene or xylene isomer) or $C_7H_6OH^+$(benzaldehyde). Beyond the identification of a few small molecules, such as acetaldehyde and methanol, interpretation of the PTR-MS measurements of complex VOC mixtures benefits from the support of another separation technique to provide identification of components.

SPME GC-MS has been combined with PTR-MS to better elucidate VOCs from plants [Aprea et al. 2009, Bouvier-Brown et al. 2007, Ruth et al. 2005] and human lungs [Bajtarevic et al. 2009, King et al. 2010]. These studies showed the combined techniques allowed for both qualitative and quantitative determinations of VOCs. The use of SPME in conjunction with PTR-MS added qualitative value to the interpretation of the PTR-MS data and strengthened the information obtained from experimentation, particularly as it applied to biological metabolic processes. The knowledge of compounds present obtained by SPME allowed elucidation of the most probable sources assigned to ions measured by the PTR-MS.

Here, PTR-MS was combined with headspace SPME GC-MS to explore the applicability of both methods for time-course resolution and quantification of volatiles emitted over the culture cycle of *A. sarcoides*. We present the results of the continuous PTR-MS method as compared with those of discrete headspace SPME GC-MS for monitoring the activity of a novel fungal culture and show both the benefits and shortcomings of each technique.

2. MATERIALS AND METHODS

2.1. Culture medium and conditions

Ascocoryne sarcoides (NRRL 50072) was provided by Dr. Gary Strobel and grown in a Biostat B Twin Plus 5-liter jacketed reactor (Sartorius Stedim Biotech) on a minimal medium consisting of (per liter): glucose (20 g), ammonium chloride (5 g), $NaH_2PO_4 \cdot 2H_2O$ (2.75 g), $MgSO_4 \cdot 7H_2O$ (0.86 g), $Ca(NO_3)_2 \cdot 4H_2O$ (.28 g), yeast extract (0.05 g), and trace salts: KCl (60 mg), KNO_3 (80 mg), $FeCl_3$ (2 mg), $MnCl_2$ (5 mg), $ZnSO_4$ (2.5 mg), H_3BO_3 (1.4 mg), and KI (0.7 mg). The minimal medium was used, in contrast to a more nutrient rich one, to minimize the amount of volatile compounds originating from the medium. An un-inoculated control reactor was run to account for background VOCs resulting from the medium constituents or air supply. Both vessels were sparged with air from compressed air cylinders at 1.5 L/min. This sparge air was humidified prior to introduction into the Biostat to minimize evaporative losses. The pH was controlled at 4.5 with 2 M HCl, 2 M NaOH and was continuously monitored by an internal Hamiltion Easyferm Plus probe. In addition, the stir rate (250 rpm) and temperature (23°C) were held constant by the Biostat control system. The reactor gaseous effluent was sent to the PTR-MS and an off-gas analyzer for O_2 and CO_2 measurements.

Accumulation of *A. sarcoides* was analyzed through daily samples of cell dry weight, where samples were filtered onto glass fiber filters (Whatman GF/F), rinsed, and dried overnight in an oven at 80°C. Once removed from the oven, filters were allowed to equilibrate at room temperature in a desiccator and then weighed. Daily glucose concentrations were measured by High Performance Liquid Chromatography (HPLC), Agilent 1200 using an Aminex HPX-87H ion exclusion column at 45°C with 0.005 M H_2SO_4 eluent.

2.2. Quantification of volatiles by PTR-MS

Proton transfer reaction-mass spectrometry (PTR-MS) was used to quantify a selected set of volatile compounds produced by *A. sarcoides*by constantly monitoring the gaseous effluent from the cultured reactor into the stationary phase of growth. The PTR-MS instrument ionized organic molecules in the gas phase through their reaction with H_3O^+, forming protonated molecules (MH^+, where M is the neutral organic molecule) and fragment ions, which were detected by a standard quadrupole mass spectrometer. This process can be used with volatiles in air with or without dilution, since the primary constituents of air (nitrogen, oxygen, argon and carbon dioxide) have a proton affinity less than water and thus are not ionized. Most organic molecules (excepting alkanes) have a proton affinity greater than water and are therefore ionized and detected [Lindinger et al. 1998].

The cultured and un-inoculated controls were analyzed by diverting a portion of the reactor effluent gas to the PTR-MS. Measurements were made from the initial inoculation until 18.5 days. At times, the gas stream was diluted with additional air to prevent premature degradation of the MS detector and keep measurements within the linear dynamic range of the PTR-MS. During dilution periods, the dilution stream was added to the reactor effluent stream and balanced with the effluent flow rate using flow controllers. Sample lines were constructed from PFA Teflon tubing and stainless steel fittings. Mass spectral scans were acquired from 20 to 220 amu at 0.5 sec/amu.

The absolute concentration of constituents in a sample were quantified directly from the measured ion intensities using the known reaction time and the theoretical reaction rate constant for the proton transfer reaction [Lindinger et al. 1998]. Concentrations reported within were derived from the PTR-MS measurements and calculated using equations derived from reaction kinetics, assuming a reaction rate coefficient of 2×10^{-9} ml s^{-1} which is appropriate for the measured compounds [Ezra et al. 2004, Lindinger et al. 1998]. This method provided a means by which the measured ion intensity at any mass can be expressed as an equivalent estimate of concentration.

The total VOC concentration produced by the fungal culture is taken as the difference between the fungal culture and the control measurements. This total reflects only the contribution from those species having proton affinities greater than that of water, which is assumed to represent the majority of the VOCs produced.

2.3. Volatile analysis with SPME GC-MS

Solid phase microextraction (SPME) was used to identify gaseous compounds as described previously [Griffin et al. 2010], but a brief description of the relevant details is provided here. Throughout the growth period, 12 mL of *A. sarcoides* culture was aseptically removed, sealed, centrifuged at 4,000 rpm for 4 minutes, then 10 mL of supernatant was added to a GC vial and stored at 4°C until analysis. For analysis, the SPME fiber (divinylbenzene/Carboxen on polydimethylsiloxane by Supelco) (Bellefonte, PA) was exposed to the vial headspace while the sample was stirred for 45 minutes. The fiber was inserted into the injection port of the GC at 240°C. The GC temperature was held at 40°C for 2 minutes, and then ramped to 230°C at 5°C min^{-1}. The mass spectrometer was tuned to meet EPA Method 8260 BFB tuning criteria. The relative amount of identified compounds was estimated by comparison with a 4-bromofluorobenzene internal standard calibration curve delivered in methanol over the concentration range of 2 to 75 μg/mL.

The quantities of volatiles desorbed from the fiber were calculated from the peak area of the total ion current measured by the mass spectrometer. This method assumed that fungal VOCs and the internal standard have the same gas-liquid partitioning, efficiency of adsorption to the fiber, and response in the GC-MS system as that of the internal standard. As will be shown later, all of these conditions are not met, and the quantities obtained are estimates. Nevertheless, the use of the internal standard allowed for a relative quantitative comparison between samples from the same culture across a time period. Data acquisition and data processing were performed on Hewlett Packard ChemStation software system. The identification of compounds produced by *A. sarcoides* was made via library comparison using the National Institute of Standards and Technology (NIST) database, and all chemical compounds described in this report use the NIST Chemistry WebBook terminology [Linstrom and Mallard 2011]. When the library identified multiple compounds as matches, careful consideration of the spectra yielded one with much higher similarity or a determination of "unknown" was made.

2.4. Henry's law constants

To characterize the headspace concentration response to a specific set of compounds at known initial concentrations using the PTR-MS and the SPME technique, a known mixture (Table 1) was introduced into the Biostat reactor system. The reactor was sparged with humidified air at 1.5 L/min, equivalent to that employed for the culturing conditions. Temperature (23°C) and stir rate (250 rpm) were held constant by the Biostat control system. A portion of the gaseous reactor effluent was diverted to the PTR-MS to measure effluent volatile concentrations over a time period of 48 hours to determine the Henry's Law constants for each volatile compound.

Liquid aliquots were removed from the reactor at 0, 1, 2, 3, 6, 12, 24, and 48 hours for analysis by headspace SPME GC-MS.

Table 1. Initial concentrations and Henry's Law constants for compounds run as standards in the abiotic experiment

		Henry's Law Constants[a]		
Standard Compound	C_o[b] (+/-)	**NIST**[c]	**Yaws**[d]	**Experimental**
Acetaldehyde	2242 (415)	14	10.3	13.9
Benzaldehyde	2239 (23)	39	39.6	38.9
Ethanol	2822 (546)	120	121.7	173
Nonanal	2032 (851)	0.7	1.5	1.2
1-Octen-3-ol	1489 (653)	62[e]	39.4[e]	25.2
1-Butanol, 3-methyl	1294 (45)		73.8	77.6

[a] *All values in mol/kg*bar*

[b] *Initial headspace concentration based on Henry's Law constants in ppbv with standard deviation in parenthesis*

[c] [Sander, R (2011)].

[d] [Yaws, CL (1999)].

[e] 1-Octanol value, value for 1-Octen-3-ol not available

The relationship governing the equilibrium concentration in the liquid in relation to the headspace is defined by Henry's Law:

$$P_A = \frac{C_{A_{liq}}}{H_A} \qquad (1)$$

Where ρ_A is the partial pressure of compound A, C_{Aliq}, is its concentration in the liquid, and H_A is the Henry's Law constant for A. Henry's Law constants were evaluated from the PTR-MS measurements following the methodology of [Karl et al. (2003)] and compared with literature values (Table 1) [Linstrom and Mallard 2011, Yaws 1999]. Experimentally determined Henry's Law constants were evaluated from the slope of the natural log of concentration versus time, and the initial substrate concentration was taken as the intercept. As the concentration of each compound decreases in the liquid, the headspace gaseous concentration in equilibrium with the liquid can be calculated from the first-order rate equation using the initial headspace concentration and the Henry's Law constant.

$$C_A(t) = C_{Ao} \exp\left[\frac{\cdot F}{H_A VRT} * t\right] \qquad (2)$$

Where C_A is the gaseous concentration of A, F is the flow rate of gas, V is the volume of liquid containing A, R is the ideal gas constant, and T is the temperature.

Initial concentration of compound A added to the reactor can be measured by PTR-MS and confirmed by calculations based on the volume of compound A added, V_A.

$$C_{Ao} = \frac{V_A P_A}{MW_A H_A PV} \qquad (3)$$

Where ρ_A is the density of A, MW_A is the molecular weight of A, and P is the pressure of the reactor headspace. The initial concentration for each standard from EQ. 3 is listed in Table 1.

3. RESULTS

3.1. Abiotic results

Abiotic results were obtained from known concentrations of six standard compounds in the reactor system under a set of controlled conditions that closely mimicked the fungal reactor system to better understand how the measurements of the PTR-MS and SPME GC/MS compare. All of the compounds were added at a level to achieve an initial headspace concentration of 2000 ppbv. The concentration profiles versus time for each of the standards as measured by the PTR-MS and the SPME fiber are shown in Figure 1. The headspace concentration of each measured standard decreased in an exponential fashion as predicted by Equation 2. Henry's Law, as an equilibrium relation, dictates that the concentration of each standard in the gas phase will decrease proportionally with time as the liquid phase concentration decreases due to sparging. The rate of decrease is determined by the air sparge rate and the system mass balance and is inversely related to the Henry's Law constant. For example, nonanal has the smallest Henry's Law constant reported in Table 1 and its headspace concentration decreased the fastest of all the analyzed compounds. Henry's law constants for each of the standards can be obtained by plotting the time series data from Figure 1 as the natural log of concentration versus time, yielding a linear relationship (not shown) where the slope of the line is related to the Henry's Law constant [Karl et al. 2003]. The Henry's Law constants calculated from the PTR data are consistent with those reported previously in the literature (Table 1), which indicates that the ions monitored are free from any interferences and the response of the PTR-MS is within the linear dynamic range of the instrument. The SPME results agree less favorably quantitatively and are discussed below. Qualitatively, the PTR-MS and SPME agree on the identification of the standard compounds with the exception of ethanol which was not detected by the SPME.

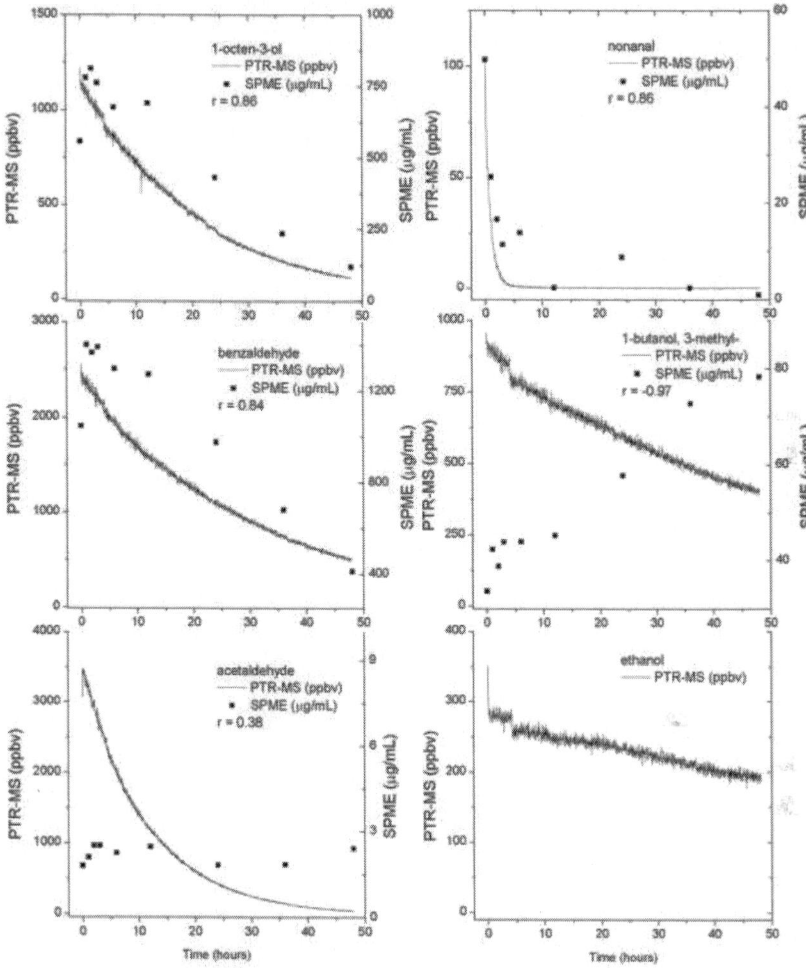

Figure 1. Abiotic concentration profiles of standard compounds as measured by PTR-MS and SPME GC-MS. Concentration profiles of each standard compound with abiotic measurements. PTR-MS response in parts per billion volume (ppbv) was monitored continuously over 48 hour time period. HS-SPME samples were removed at 0, 1, 2, 3, 6, 12, 24, 36 and 48 hours with correlation coefficients corresponding to six minute PTR-MS averages.

The SPME results in Figure 1 show that the divinylbenzene/Carboxen on polydimethylsiloxane fiber is inefficient at trapping acetaldehyde and ethanol. This result is not totally unexpected, as the fiber is not specified for the analysis of C2 compounds. It appears that the SPME fiber did not provide results proportional to the amount of acetaldehyde or ethanol present. Additionally, the SPME measurements of 1-butanol, 3-methyl- had a contradictory trend to the PTR-MS measurements and did not decrease in concentration over time. The SPME results for the higher molecular weight components appear to more closely follow the anticipated Henry's law behavior and are fairly consistent with the PTR-MS results with correlation coefficients above 0.8.

A better means of accessing the correlation between the PTR-MS and SPME data can be made by collecting the concentration data at the same discrete time points. The PTR-MS and SPME time point concentrations have been plotted against each other in Figure 2 with the corresponding correlation factors. Note that ethanol was not included in Figure 2, as there was no ethanol peak detected in the SPME GC-MS technique. While it is apparent in both Figures 1 and 2 that the measurement of acetaldehyde and 1-butanol, 3-methyl- by the two techniques do not correlate well, Figure 2 shows that the degree of correlation for benzaldehyde and 1-octen-3-ol is concentration dependent. The PTR-MS and SPME measurements appear to track one another at the lower concentrations. At the higher concentrations, the SPME results appear to saturate, which suggests that the adsorption capacity of the fiber may have been exceeded. The nonanal results appear to be similarly affected, but the comparison is less definitive due to rapid removal of nonanal from solution and the appearance of the non-zero intercept. The contradictory trend for 1-butanol, 3-methyl- suggests that the adsorption process is competitive, because the 1-butanol, 3-methyl-concentration, as measured by SPME, increases as the stronger adsorbing species decrease in concentration as they are removed from solution. These results indicate that the concentrations measured by the SPME technique are strongly dependent not only on the type of mixture constituents, but also on the relative concentrations of those components. Thus even in relatively simple systems such as this, some care must be exercised in deducing quantitative information from the SPME data.

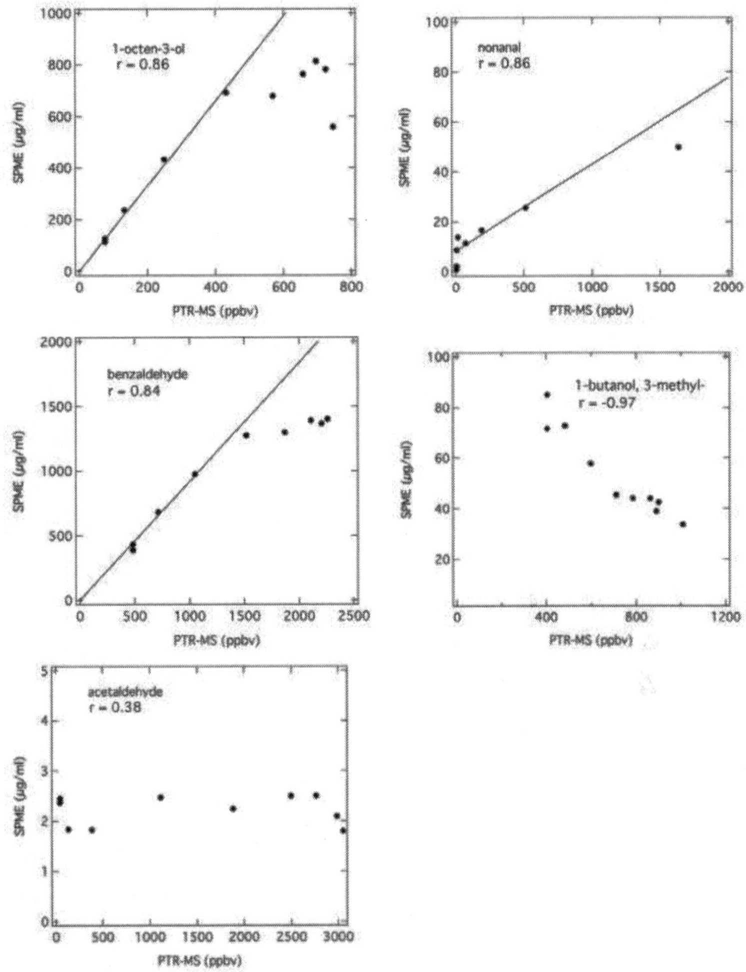

Figure 2. Abiotic correlation of standard compound concentrations. Correlation of relative concentration measurements for PTR-MS and the SPME GC-MS technique for all compounds with abiotic measurements except for ethanol, which had a non-detect response by SPME GC-MS. PTR-MS values are centered at SPME sample collection times and are six minute averages. Lines indicate linear correlation at low concentrations with divergence at high concentrations while correlation coefficients include all data points.

3.2. Quantification of volatile products of *A. sarcoides*

Applying these techniques to the biological system, Figure 3 shows the time series of the total VOC signal (sum of all the ions from m/z 40-205) as measured by the PTR-MS along with a measure of *A. sarcoides* biological growth expressed as biomass density. The initial variation in the VOC signal is due to inoculation of the reactor with an active culture, and the initial peak decays away due to the sparging process. The culture demonstrated a lag phase until day 3 before entering an exponential growth phase. The total VOC concentration increased as the fungus entered exponential phase indicating growth associated product synthesis. The flow rate of gas supplied to the reactor was constant such that increasing concentrations demonstrates increased metabolic production of VOCs. The maximum ion concentration of VOCs measured was 10,000 ppbv at 284 hours (near 12 days). The cell concentration began to stabilize on day 12 around 4.5 g/L indicating the culture's entry into stationary phase. Once the stationary phase was reached after day 12, the total VOC signal stabilized, and then dropped rapidly around 380 hours. Throughout stationary phase, the PTR-MS measured the VOCs excluding acetaldehyde and ethanol at 2,000 ppbv. The very sharp transitions in the VOC signal are due to variations in the sparge flow and should not be interpreted as specific metabolic changes.

Continuous monitoring by the PTR-MS revealed that two ions (m/z 45 (acetaldehyde) and m/z 47 (ethanol)) comprised the majority of the ion signal from the *A. sarcoides* culture. Excluding m/z 45 and m/z 47, the remaining ion intensity measured was between 1,000 and 2,000 ppbv in concentration for the majority of the time (Figure 3). The remaining ions account for ~ 14% of the total VOC signal. A listing of all of the ions whose intensity exceeded 1 ppbv is tabulated in Table 2 along with their signal intensity at discrete time points. Ions of the carbon 13 isotopes are not included, i.e. m/z 72 which is 5.5% of m/z 71. Total production of compounds in Table 2 increased to over 2,000 ppbv at 280 hours and leveled off as the fungus entered stationary phase after day 12. The majority of these compounds were composed of 4 and 5 carbon alcohols, acetic acid or acetate esters and compounds producing m/z 59 (acetone and/or propanal). The PTR-MS signal from 1-butanol, 3-methyl- was over 1% of the total VOCs. Notable compounds of fuel interest included 1-octen-3-ol, nonanal, 1-butanol, 3-methyl- and benzaldehyde (Figure4). The PTR-MS signals of the alcohols in Figure 4 increased steadily until near the end of the growth cycle with a peak around 275 hours at the end of the exponential growth phase. The benzaldehyde signal reached a maximum around 210 hours before halving just before 300 hours and remained there for the stationary growth phase of *A. sarcoides*. The SPME measured concentration profiles correspond well with the PTR-MS results for the higher molecular weight alcohols (correlation coefficients > 0.8), but not for ethanol or benzaldehyde (correlation coefficients < 0.7). An increase in the concentration of a compound should elicit a relative increase in the measured response of each technique. Figure 4 and Table 2 clearly show that there are compounds for which the SPME response to a change in VOC concentration does not scale with the corresponding PTR-MS signal change. The differences in response show the confounding effects of SPME fiber selectivity.

Table 2. Select ion signals in the dynamic headspace of *A.sarcoides* culture measured by PTR-MS

Mass (amu)[a,b]	**Signal Intensity (ppbv) at given age in days**[c]						
	2	**5**	**9**	**11**	**13**	**15**	**17**
33 [methanol]	1.0	3.6	6.8	9.1	11.1	10.6	12.8
41	6.3	12.7	27.1	40.8	48.8	49.7	52.2
43	36.6	87.8	198	264	251	260	269
57	11.1	44.7	129	180	187	201	234

Mass (amu)[a,b]	Signal Intensity (ppbv) at given age in days[c]						
	2	5	9	11	13	15	17
59 [C_3H_6O]	17.0	71.9	198	264	251	260	269
61	7.6	23.2	69.6	87.7	60.4	54.8	52.5
69	1.7	3.5	5.9	7.5	6.6	7.3	8.6
71 [1-butanol, 3-methyl-]	12.9	24.0	76.4	99.1	98.6	111	127
73 [C_4H_8O]	4.4	39.8	38.6	41.3	43.8	44.1	38.2
75	0.2	0.5	1.9	2.3	1.4	1.5	1.6
87 [$C_5H_{10}O$]	1.0	2.2	4.0	4.9	7.4	10.0	8.0
89 [ethyl acetate]	5.3	14.1	47.9	53.4	32.0	31.0	31.6
91	0.3	0.8	2.0	2.6	2.0	2.1	2.0
93	0.2	0.2	0.9	1.1	0.6	0.8	0.9
105	0.2	0.5	1.8	2.4	2.1	2.2	2.6
107 [benzaldehyde]	0.2	6.3	7.5	6.9	3.5	3.4	3.3
111 [1-octen-3-ol]	0.1	0.5	2.5	3.4	2.6	3.8	4.4
115	0.0	0.5	1.8	1.8	0.9	0.5	0.7
117	0.2	0.4	1.5	2.1	2.0	3.0	4.2

Mass (amu)[a,b]	Signal Intensity (ppbv) at given age in days[c]						
	2	5	9	11	13	15	17
129	0.1	0.4	0.9	1.3	1.2	1.2	1.5
143 [nonanal]	0.1	0.6	0.6	0.7	0.2	0.4	0.5
SUM	106	338	847	1086	1013	1063	1153

[a] Acetaldehyde (45), ethanol (47) and minor ion signals are excluded.

[b] Tentative assignments to the ion are given in brackets.

[c] Signal intensity denoted as 4 hour averages associated with the data in Table 3.

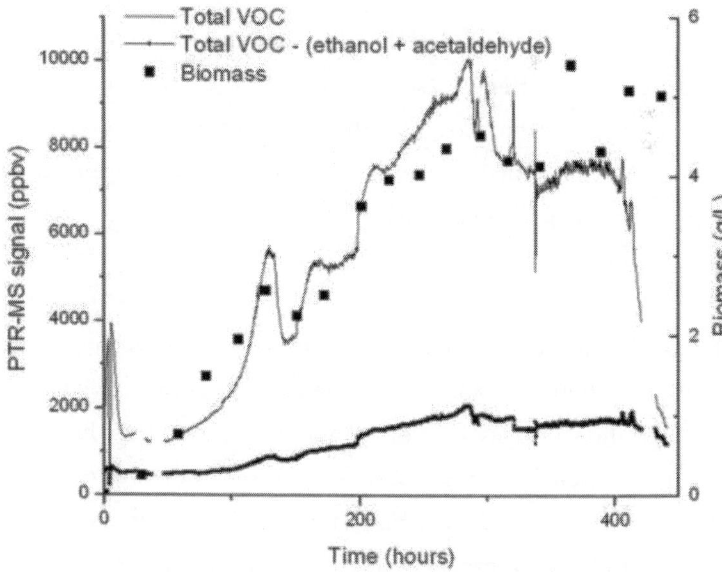

Figure 3. Total VOCs in sparged air from *A. sarcoides* culture as measured by PTR-MS.The total VOCs excluding the signals from ethanol and acetaldehyde were near 2000 ppbv at their maximum. Biomass was measured by dry weight.

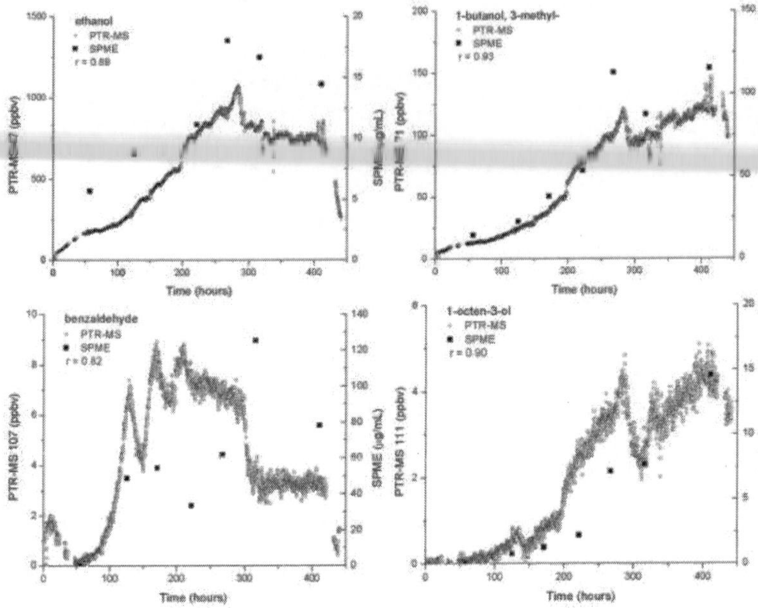

Figure 4. Measured concentrations of ethanol, 1-butanol, 3-methyl-, benzaldehyde and 1-octen-3-ol over the course of the growth period. Response of ethanol, 1-butanol, 3-methyl-, benzaldehyde and 1-octen-3-ol over the course of the growth period as measured by PTR-MS and HS-SPME GC-MS including correlation coefficient, r.

The total ion signal was consistent with *A. sarcoides* biomass concentration; meaning VOC production was closely related with growth (Figure 3). The VOC concentration profiles vary by compound as shown in Figure 4 with some being associated with growth. The compounds with the highest concentrations were acetaldehyde and ethanol. The major ion signatures, m/z 101, m/z 119, and m/z 159 found in the [Strobel et al. (2008)] experiment on oatmeal agar were not observed in this study. This change in product composition is most likely due to different carbon source and experimental conditions.

3.3. Identification of volatile products of *A. sarcoides*

The qualitative information provided by the SPME technique identifies specific VOCs, thereby allowing better ion assignments to be made with the PTR-MS. Most of the compounds positively identified by the SPME technique were also detected with PTR-MS, though the relative amounts of the identified compounds differed significantly between the two techniques. In some instances, the ions detected by PTR-MS were not unique and reflect the collective signal from more than one compound; often the case for lower mass ions. A total of twenty-seven distinct peaks were found using SPME GC-MS, with eight being unidentifiable (Table 3). Of the identified compounds, there was a mixture of (in order of descending frequency) organic acids and esters, alcohols, aldehydes, aromatics and alkanes. Two alkanes were identified by the SPME GC-MS (pentane and 4-methyl-heptane). The branched 8 carbon alkane, 4-methyl-heptane, was produced from Day 2 to Day 15. Physical property information for Table 3 compounds is listed in Additional file 1.

Table 3. GC-MS head-space analysis of VOCs produced by *A.sarcoides* culture collected with a SPME fiber

Retention	**Possible Compound**[a]		**Peak Area at given age in days**[b]			
Time (min)	**(Molecular mass (Da))**	**2**	**5**	**9**	**15**	**17**
1.34	Unknown (56)	17.8	13.0	ND	ND	ND
1.38	Pentane (72)[c]	ND	12.6	38.7	22.6	31.8
2.01	Heptane, 4-methyl-	12.9	12.6	11.1	15.6	ND

Retention	Possible Compound[a]		Peak Area at given age in days[b]			
Time (min)	(Molecular mass (Da))	2	5	9	15	17
	(114)[d]					
2.09	Unknown (118)	10.2	9.1	7.3	8.8	6.7
3.26	Ethyl acetate (88)[c]	10.1	19.5	52.9	51.8	44.8
3.89	Ethanol (46)[d]	79.3	121.8	157.3	223.5	188.2
6.07	Unknown (114)	ND	16.1	27.5	45.2	ND
6.40	1-Propanol, 2-methyl- (74)[c,d]	10.1	20.1	17.7	74.9	52.8
7.70	Unknown	11.7	11.0	7.7	11.0	23.7
8.19	1-Butanol, 3-methyl- (impure) (88)[c,d]	88.6	139.7	334.1	665.5	667.6
9.15	Unknown (134)	8.0	10.7	12.8	ND	ND
9.56	Unknown (159 or 88)	ND	11.7	18.7	35.8	ND
11.65	1-Octen-3-ol (128)[c,d(isomer)]	ND	7.6	22.8	80.3	84.5
12.02	Acetic acid, octyl ester (172)[c]	15.6	7.8	3.4	2.6	4.4

Retention	Possible Compound[a]		Peak Area at given age in days[b]			
Time (min)	(Molecular mass (Da))	2	5	9	15	17
12.95	Benzaldehyde (106)	5.2	302.8	211.3	679.7	453.5
13.33	Acetic acid, nonyl ester (186)[c]	33.9	32.8	7.5	5.5	ND
13.99	Unknown (122)	10.9	32.8	1.7	ND	ND
14.58	Acetic acid, n-decyl ester (200)[c,d]	19.3	9.1	ND	ND	ND
14.86	$C_{15}H_{24}$, sesquiterpene (204)[c]	19.3	8.1	5.4	4.1	ND
15.95	Nonanal (142)	15.1	12.1	12.3	17.6	46.7
16.99	Benzyl alcohol (108)[c,d]	ND	5.1	39.3	39.1	34.9
18.37	Unknown (126)	4.4	5.1	10.0	17.0	26.8
18.71	Decanal (156)	ND	ND	ND	24.7	60.8
23.27	Butanoic acid, 3-methyl- (102)[c]	ND	ND	ND	8.9	20.3
24.83	Oxime-, methoxy-phenyl- (151)	ND	ND	ND	39.8	35.7
26.30	Acetic acid, 2-phenylethyl ester	ND	ND	ND	24.5	24.0

Retention Time (min)	Possible Compound[a] (Molecular mass (Da))	Peak Area at given age in days[b]				
		3	5	9	15	17
	(164)[c]					
28.33	Phenylethyl alcohol (122)[c,d]	ND	ND	ND	37.6	36.6

[a] Unknown compounds represent those whose assignment is uncertain. However, probable molecular masses are included when consistent.

[b] Peak areas listed as 10^6; ND = signal not detected

[c] identified by Griffin et al. 2010 (12)

[d] identified by Strobel et al. 2008 (32, 33)

As noted in Table 3, twelve compounds identified in this study were also detected by [Griffin et al. (2010)] on various media including cellulose. Although there were major differences in the volatile constituents identified, this was not surprising due to the very different experimental conditions of this study. For example, acetic acid, heptyl ester was reported [Strobel et al. 2008] as one of most abundant compounds, but it was not identified by SPME in these results. However, many minor compounds identified by [Strobel et al. (2008)] on oatmeal agar were identified in this study despite conservative identification changes later made [Strobel et al. 2010]. These compounds are noted in Table 3 and include heptane, 2-methyl-; acetic acid, decyl ester; ethanol; 1-propanol, 2-methyl-; phenylethyl alcohol; 1-butanol, 3-methyl-; and an isomer of 3-octen-2-ol (1-octen-3-ol).

Of the compounds identified by SPME GC-MS, the concentration profiles of several correlated with the PTR-MS measured concentrations with correlation coefficient exceeding 0.80 (Tables 2 and 3). These compounds include acetic acid, ethyl ester (89) and the alcohols: 1-butanol, 3-methyl-

(71), 1-octen-3-ol (111) (Figure 4). The concentration profile of 1-butanol, 3-methyl- as indicated by SPME correlated well with the PTR-MS measurements of ion 71 (Figure 4) with a correlation coefficient of 0.93, despite the negative correlation found by the abiotic results (Figures 1 and 2). A possible explanation for this is the concentrations of other compounds in the fungal gas phase were small and did not preferentially displace 1-butanol, 3-methyl- from the fiber, while in the abiotic results, this was not the case. The concentration profiles for 1-octen-3-ol were strikingly similar despite the relatively low concentration of this product. This was also the case for butanoic acid, 3-methyl- and nonanal whose PTR-MS measured concentrations were less than 1 ppbv. Like 1-butanol, 3-methyl-, their measured concentrations steadily increased through the majority of the time points. The variance in reported signals is seen in Tables 2 and 3 and Figure 4.

The SPME measurements of compounds that did not correlate well with the PTR-MS signal included benzaldehyde (107) and 1-propanol, 2-methyl- (57). For benzaldehyde, the first SPME data point matches the PTR-MS signal, but afterwards, the measured SPME concentrations are out of step with PTR-MS concentrations (Figure 4). There was a similar incongruity with the 1-propanol, 2-methyl- measurements. The PTR-MS measurement of 1-propanol, 2-methyl- is made by monitoring the m/z 57 ion. The m/z 57 ion, however, is not unique to 1-propanol, 2-methyl-, and it is probable that the lack of agreement between the PTR-MS and SPME results may be due to signal interference in the PTR-MS measurement. While perturbations in the PTR-MS signal for benzaldehyde could be similarly explained by signal interference, but the abiotic results for benzaldehyde showed no interference, eliminating this possibility.

4. DISCUSSION

Taken together, the results of the PTR-MS and SPME GC-MS techniques provide a more definitive description of the VOCs produced by *A. sarcoides* for product (e.g. biofuel) development. However, the challenges associated with the quantitative and qualitative analysis of complex VOC mixtures in an aqueous solution are non-trivial, and neither method had the ability to both identify and quantify the breadth of compound diversity synthesized by this complex fungal system.

The *A. sarcoides* culture consistently produced numerous VOCs. There was little quantitative agreement in VOC concentrations between the SPME technique and the PTR-MS for the suite of compounds produced by *A. sarcoides*. Both techniques have advantages for exploring potential fuel products from fungal or other cultures. SPME effectively identifies a range of compounds produced by the culture, and at modest concentrations, has adequate fiber capacity for many relevant compounds. The use of PTR-MS gives quantitative data on the evolution of volatiles permitting temporal resolution of microbial volatile production. These data can be used to help elucidate real-time metabolism shifts of an organism for product applications (excluding alkanes) and metabolic modeling. The PTR-MS results may also be valuable in guiding genetic engineering efforts to improve production rates of specific compounds..

To effectively use PTR-MS and SPME GC-MS, the inherent differences in signal acquistion must be considered. Foremost, PTR-MS is a continuous measurement of VOCs in the gaseous stream stripped from an aqueous culture; SPME is a measurement of the headspace compounds that develop above a liquid aliquot after a 45 minute exposure and equilibration period. However, SPME is not truly an equilibrium measurement, because the concentration of the headspace is potentially being depleted by compounds adsorbing to the fiber, lowering their vapor pressure and thereby constantly shifting the headspace composition (Eq. 1) [Zeng and Noblet 2002]. SPME adsorbs VOCs from the headspace then estimates a liquid concentration based on one internal standard. Therefore, we anticipate differences in distribution of VOCs detected by the two methods (Figures 1 and 2). The measurement variability that exists between the two techniques should decrease if the gaseous effluent measured by the PTR-MS is at or near equilibrium with the compounds in the liquid. Based on the abiotic experimental results with a known mixture of VOC standards, equilibrium was assumed for the experimental reactor system.

A limitation of PTR-MS is that alkanes are not ionized by H_3O^+ and therefore are not detectable making it unsuitable for some biofuel applications. Additional quantitative complications can arise due to the presence of fragment ions (decomposition products of the protonated molecule). While ionization via proton transfer with H_3O^+ is considered to be a relatively soft ionization method, many of the components produced by *A. sarcoides* fragment extensively. Most notable are the alcohols that, except for methanol, fragment by the elimination of H_2O from the protonated molecule [Buhr et al. 2002]. Thus, the alcohols are detected as an ion corresponding to $(M\text{-}OH)^+$. Despite these considerations, the results clearly demonstrate the applicability of the PTR-MS for time series resolution of volatiles from biological cultures.

SPME analysis should reflect the chemical composition and capacity on the SPME fiber if possible. However, not all compounds adsorb to the fiber material with equal affinities or at equal rates, such that the use of one internal standard is not sufficient to determine all constituent concentrations [Pawliszyn 1997]. The preferential adsorption by some compounds appears to be further exacerbated if the capacity of the fiber is exceeded. The least volatile, highest affinity compounds will be extracted from the headspace-liquid matrix last, therefore adequate extraction times must be observed [Pawliszyn 1999]. Therefore, drawing conclusions about the concentration of constituents in complex mixtures is very difficult due to fiber selectivity and saturation characteristics. In general, the SPME fiber appeared to preferentially adsorb the higher molecular weight compounds and therefore over predicts their abundance (Table 3). While these ions may have been detected by the PTR-MS, signals were low in concentration either because their headspace concentrations were low or because of possible adsorptive losses in the sample lines and therefore were not included in Table 2. An outcome of this preferential adsorption was seen in the 1-butanol, 3-methyl- data. In the fungal culture, the SPME data correlated well with the PTR-MS signal (Figure 4), but not in the abiotic data (Figure 1) because the culture headspace environment was dominated by lower molecular weight compounds. The negative trend seen in the abiotic results was most likely because the higher molecular weight compounds were preferentially adsorbed by the SPME fiber, leaving few adsorption sites for the other compounds (Figure 1). The extent of fiber bias for the lower volatility, higher molecular weight compounds is not easily predicted or measured for complex mixtures. However, SPME appears to be more applicable in the fungal headspace environment than in conditions where higher concentrations of higher molecular weight compounds were present, such as in the abiotic system.

The PTR-MS and SPME results demonstrated the quantitative limits of the SPME technique even when an internal standard was used. However, use of SPME to estimate relative concentrations of higher molecular weight compounds may be reasonable if application is restricted to qualitative temporal trends and not individual quantitative time points. A possibility for improving the SPME technique is to use replicate samples with additional fiber types to better capture and characterize the breadth of compounds present.

This study shows that despite their shortcomings, the combined use of PTR-MS with a qualitative technique such as GC-MS by headspace SPME is a valuable method for obtaining more detailed analysis of the VOCs produced by microbial systems and can be applied to biofuel product development. However, when quantifying compounds as produced by metabolic processes, especially when those compounds are in a complex mixture, the combination of multiple techniques is warranted until a single more robust qualitative and quantitative analytical method is developed.

ACKNOWLEDGEMENTS

The authors would like to thank Elle Pankratz for her general laboratory assistance and continuous hard work. This work was supported by the National Science Foundation (NSF) Emerging Frontiers in Research and Innovation (EFRI), Grant No. 0937613 and Chemical, Bioengineering, Environmental and Transport Systems (CBET), Grant No. 0802666. Additional funding for Natasha Mallette was provided by the NSF Integrated Graduate and Education Research Traineeship (IGERT) Program.

Manuscript preparation was supported by the National Science Foundation (NSF) Emerging Frontiers in Research and Innovation (EFRI), Grant No. 0937613

REFERENCES

1. Aprea E, Biasioli F, Carlin S, Endrizzi I, Gasperi F: **Investigation of Volatile Compounds in Two Raspberry Cultivars by Two Headspace Techniques: Solid-Phase Microextraction/Gas Chromatography Mass Spectrometry (SPME/GC MS) and Proton-Transfer Reaction Mass Spectrometry (PTR MS).** *J Agr Food Chem* 2009, **57:**4011–4018.

2. Aprea E, Biasioli F, Carlin S, Versini G, Märk TD, Gasperi F: **Rapid white truffle headspace analysis by proton transfer reaction mass spectrometry and comparison with solid-phase microextraction coupled with gas chromatography/mass spectrometry.** *Rapid Commun Mass Sp* 2007, **21:**2564–2572.

3. Arthur CL, Pawliszyn J: **Solid phase microextraction with thermal desorption using fused silica optical fibers.** *Anal Chem* 1990,**62:**2145–2148.

4. Bäck J, Aaltonen H, Hellén H, Kajos MK, Patokoski J, Taipale R, Pumpanen J, Heinonsalo J: **Variable emissions of microbial volatile organic compounds (MVOCs) from root-associated fungi isolated from Scots pine.** *Atmos Environ* 2010, **44:**3651–3659.

5. Bajtarevic A, Ager C, Pienz M, Klieber M, Schwarz K, Ligor M, Ligor T, Filipiak W, Denz H, Fiegl M, Hilbe W, Weiss W, Lukas P, Jamnig H, Hackl M, Haidenberger A, Buszewski B, Miekisch W, Schubert J, Amann A: **Noninvasive detection of lung cancer by analysis of exhaled breath.** *BMC Cancer* 2009, **9:**348.

6. Bouvier-Brown NC, Holzinger R, Palitzsch K, Goldstein AH: **Quantifying sesquiterpene and oxygenated terpene emissions**

from live vegetation using solid-phase microextraction fibers. *J Chromatogr A* 2007, **1161:**113–120.

7. Brás I, Ratola N, Alves A: **Uncertainty in the quantification of pentachlorophenol in wood processing wastewaters by SPME GC-MS.** *J Anal Chem* 2011, **66:**756–762.

8. Buhr K, van Ruth S, Delahunty C: **Analysis of volatile flavour compounds by Proton Transfer Reaction-Mass Spectrometry: fragmentation patterns and discrimination between isobaric and isomeric compounds.** *Int J Mass Spectrom* 2002, **221:**1–7.

9. Bunge M, Araghipour N, Mikoviny T, Dunkl J, Schnitzhofer R, Hansel A, Schinner F, Wisthaler A, Margesin R, Mark TD: **On-Line Monitoring of Microbial Volatile Metabolites by Proton Transfer Reaction-Mass Spectrometry.** *Appl Environ Microbiol* 2008,**74:**2179–2186.

10. Ezra D, Jasper J, Rogers T, Knighton B, Grimsrud E, Strobel G: **Proton transfer reaction-mass spectrometry as a technique to measure volatile emissions of Muscodor albus.** *Plant Sci* 2004, **166:**1471–1477.

11. Griffin MA, Spakowicz DJ, Gianoulis TA, Strobel SA: **Volatile organic compound production by organisms in the genus Ascocoryne and a re-evaluation of myco-diesel production by NRRL 50072.** *Microbiology* 2010, **156:**3814–3829.

12. Kai M, Crespo E, Cristescu SM, Harren FJM, Francke W, Piechulla B: **Serratia odorifera: analysis of volatile emission and biological impact of volatile compounds on Arabidopsis thaliana.** *Appl Microbiol Biotechnol* 2010, **88:**965–976.

13. Karl T, Yeretzian C, Jordan A, Lindinger W: **Dynamic measurements of partition coefficients using proton-transfer-reaction mass spectrometry (PTR-MS).** *Int J Mass Spectrom* 2003, **223–224:**383–395.

14. King J, Mochalski P, Kupferthaler A, Unterkofler K, Koc H, Filipiak W, Teschl S, Hinterhuber H, Amann A: **Dynamic profiles of volatile organic compounds in exhaled breath as determined by a coupled PTR-MS/GC-MS study.** *Physiol Meas* 2010, **31:**1169–1184.

15. Ladygina N, Dedyukhina EG, Vainshtein MB: **A review on microbial synthesis of hydrocarbons.** *Process Biochem* 2006, **41:**1001–1014.

16. Langenfeld JJ, Hawthorne SB, Miller DJ: **Quantitative Analysis of Fuel-Related Hydrocarbons in Surface Water and Wastewater Samples by Solid-Phase Microextraction.** *Anal Chem* 1996, **68:**144–155.

17. Lindinger W, Hansel A, Jordan A: **On-line monitoring of volatile organic compounds at pptv levels by means of proton-transfer-reaction mass spectrometry (PTR-MS) medical applications, food control and environmental research.** *Int J Mass Spectrom* 1998,**173:**191–241.

18. Linstrom PJ, Mallard WG (Eds): **NIST Chemistry WebBook, NIST Standard Reference Database Number 69** In National Institute of Standards and Technology, Gaithersburg MD, 20899; 2011. Accessed 3 Oct 2011

19. Maleknia S, Bell T, Adams M: **PTR-MS analysis of reference and plant-emitted volatile organic compounds.** *Int J Mass Spectrom*2007, **262:**203–210.

20. Mallouchos A, Komaitis M, Koutinas A, Kanellaki M: **Investigation of Volatiles Evolution during the Alcoholic Fermentation of Grape Must Using Free and Immobilized Cells with the Help of Solid Phase Microextraction (SPME) Headspace Sampling.** *J Agr Food Chem* 2002, **50:**3840–3848.

21. Parkerton TF, Stone MA, Letinski DJ: **Assessing the aquatic toxicity of complex hydrocarbon mixtures using solid phase microextraction.** *Toxicol Lett* 2000, **112–113:**273–282.

22. Pawliszyn J (Ed). *Applications of solid phase microextraction* The Royal Society of Chemistry, Cambridge; 1999.

23. Pawliszyn J: *Solid Phase Microextraction: Theory and Practice.* Wiley-VCH, New York; 1997.

24. Ramirez KS, Lauber CL, Fierer N: **Microbial consumption and production of volatile organic compounds at the soil-litter interface.***Biogeochemistry* 2009, **99:**97–107.

25. Robinson AL, Boss PK, Heymann H, Solomon PS, Trengove RD: **Development of a sensitive non-targeted method for characterizing the wine volatile profile using headspace solid-phase microextraction comprehensive two-dimensional gas chromatography time-of-flight mass spectrometry.** *J Chromatogr A* 2011, **1218:**504–517.

26. Ruth V, M S, Lonneke D, Eugenio A, Sachiko O: **Comparison of Volatile Flavour Profiles of Kidney Beans and Soybeans by GC-MS and PTR-MS.** *Food Sci Technol Res* 2005, **11:**63–70.

27. Sander R: **"Henry's Law Constants" in NIST Chemistry WebBook, NIST Standard Reference Database Number 69.** In Edited by: Linstrom P.J, Mallard W.G. National Institute of Standards and Technology, Gaithersburg MD, 20899; 2011.

28. Solomons TWG, Fryhle CB: *Organic Chemistry [Book].* Wiley, New York, New York; 2002.

29. Steeghs M: **Proton-Transfer-Reaction Mass Spectrometry as a New Tool for Real Time Analysis of Root-Secreted Volatile Organic Compounds in Arabidopsis.** *Plant Physiol* 2004, **135:**47–58.

30. Strobel GA, Knighton B, Kluck K, Ren Y, Livinghouse T, Griffin M, Spakowicz D, Sears J: **The production of myco-diesel hydrocarbons and their derivatives by the endophytic fungus Gliocladium roseum (NRRL 50072).** *Microbiology* 2008, **154:**3319–3328.

31. Strobel GA, Knighton B, Kluck K, Ren Y, Livinghouse T, Griffin M, Spakowicz D, Sears J: **The production of myco-diesel hydrocarbons and their derivatives by the endophytic fungus Gliocladium roseum (NRRL 50072).** *Microbiology* 2010, **156:**3830–3833.

32. Tomsheck AR, Strobel GA, Booth E, Geary B, Spakowicz D, Knighton B, Floerchinger C, Sears J, Liarzi O, Ezra D: **Hypoxylon sp., an Endophyte of Persea indica, Producing 1,8-Cineole and Other Bioactive Volatiles with Fuel Potential.** *Microb Ecol* 2010, **60:**903–914.

33. Weldegergis BT, Crouch AM, Górecki T, de Villiers A: **Solid phase extraction in combination with comprehensive two-dimensional gas chromatography coupled to time-of-flight mass spectrometry for the detailed investigation of volatiles in South African red wines.** *Anal Chim Acta* 2011, **701:**98–111.

34. Wercinski SAS (Ed): *Solid phase microextraction - a practical guide* CRC Press; 1999.

35. Yaws CL: *Chemical Properties Handbook.* McGraw-Hill; 1999.

36. Zeng EY, Noblet JA: **Theoretical Considerations on the Use of Solid-Phase Microextraction with Complex Environmental Samples.** *Environ Sci Technol* 2002, **36:**3385–3392

CHAPTER 5

Characterization and Identification of Natural Terpenic Resins employed in "*Madonna con Bambino e Angeli*" by *Antonello da Messina* using Gas Chromatography–Mass Spectrometry

Mario Vincenzo Russo[1] and Pasquale Avino[1,2]

[1]*Facoltà di Agraria (DISTAAM), Università del Molise, via de Sanctis, Campobasso 86100, Italy.*
[2]*DIPIA, INAIL ex-ISPESL, via Urbana 167, Rome I-00184, Italy.*

ABSTRACT

Background

Natural resins were frequently employed as adhesives or as components of oleo-resinous media in paintings in the past. The identification of vegetable resins is still an open problem. The aim of this paper is to analyze by GC-MS some vegetable resins frequently employed in paintings, such as Venice turpentine, dammar, copal, elemi in order to identify their main component in raw and aged samples. Some molecules are proposed as chemical "*markers*" to identify these natural resins.

Results

The results obtained on standards allowed us to successfully analyze sample collected from one work of art: the *Madonna with the Infant and Angels* by Antonello da Messina (XV century).

Conclusion

The results obtained confirm that the painting the artist originally used a mixture of linseed oil and natural resin (Venice turpentine) as binding medium.

KEYWORDS

Binding media Paintings Natural resins Marker GC-MS Madonna with the Infant and Angels Antonello da Messina.

Background

The identification of the compounds employed in artistic paintings provides valuable information for both the knowledge of an artist and a correct restoration project. In particular, it is the analysis of binding media that allows it; simultaneously, this issue has always been one of the most important goals of analytical chemistry in conservation studies [1–4]. Identification of binding media involves several analytical problems mainly related to the very low amount of sample available and its heterogeneity, and very low content of organic media components as well as unknown state of alteration/degradation.

A difficult but challenging task in these studies is the identification of markers. These are characteristic compounds that allow the precise identification of a resin or of a drying oil in a sample taken from a painting. Indeed, to reveal the presence of a particular substance used by an artist, it is necessary to find out suitable markers.

Table 1. Characteristic fragment ions and corresponding m/z values of dehydroabietic acid (DHA) and 7-oxo-dehydroabietic acid (7-oxo-DHA) as methyl esters

Ion	DHA	7-oxo-DHA
$M^{+\cdot}$	314	328
$[M - CH_3]^+$	299	313
$[M - COOCH_3]^+$	255	269
$[M - HCOOCH_3]^+$		268
$[M - CH_3HCOOCH_3]^+$	239	253

Binding media are natural products of relatively complex nature that are used as film-forming substances to give cohesion to pigments, to adhere them to each other and to a backing substrate and to protect them from deterioration [5–10]. Analytical chemistry offers suitable techniques to provide the restorer with the necessary analytical information for conservation treatment [11].

Binding media are suitable for a classification based on their chemical nature. Linseed oil is the most widely employed drying oil. Among proteinaceous substances, the most commonly used are: egg yolk, egg white and casein. Glues of animal origin have often be seen employed, both by artists and restorers: they are mainly constituted from collagen and keratin. Among the proteinaceous glues, beef and porcine gelatines, albumin, casein and egg protein are the most widely employed as binding media in tempera paintings [12–16].

Natural resins of vegetal origin were widely employed in easel painting [17]. Different natural resins were mainly used for decorative and varnish purposes since very ancient times for their excellent glasslike capability. A great number of recipes and preparation procedures are documented in literature [18, 19]. Artists used them as final varnish for their protective properties and with the aim of improving optical properties of painting films. Sometimes resins were added to drying oils to modify the fluidity of the binding media. Resins form a protective coat over a paint film and give an uniform surface to the work. Besides, they improve the optical qualities of the picture by increasing both the saturation of the color and the overall gloss.

Among the various class of organic materials used as binding media, the natural resins are those which still have the reputation of being difficult to identify and suffer from analytical neglect [12, 16]. Natural resins are secreted by a large variety of plants. They are largely composed by mono-, sesqui-, di- and triterpenoids. Interesting and very useful for classification and identification, is the fact that di-and triterpenoids are not found together in the same resin [15]. Both artists and restorers have largely used, in virtually all parts of paintings, diterpenic resins. Venice turpentine and Strasbourg turpentine were most popular. Pine resins, first of all colophony, presented the problem of becoming dark and brittle. Nevertheless sources suggest that pine colophony-drying oil mixtures where frequently used [20]. Triterpenoids can be either tetracyclic or pentacyclic. The tetracyclic compounds are based either on the dammarane skeleton or the euphane skeleton. The most popular triterpenic resins used in painting are dammar and mastic [19, 21–23].

This paper focuses on the application of chromatographic techniques to the preservation and conservation of one of the most important paintings of the world cultural heritage. Gas chromatography–mass spectrometry (GC-MS) was employed, as it is a well-established technique for the analysis of complex mixtures and holds a prime position in analytical chemistry because of its combination of sensitivity, wide range of applicability and versatility [24]. In particular, a painting of the XV century, the *Madonna with the Infant and Angels* ("Madonna con Bambino e Angeli") by Antonello da Messina, was studied and the composition of the natural resins of the binding media was investigated. From the artistic point of view, Antonello da Messina was the first Italian artist to use the oil painting in Italy due his strong relationship with Flemish masters.

In particular, drying oils and natural resins of vegetal origin were studied in order to get information about ageing processes that consist in modification and degradation of these compounds. The study was focused on drying oils and varnishes. The aim was to determine the drying oils and the resins originally employed by the artist and by restorers. Micro Fourier Transform-Infrared spectroscopy (FT-IR) analyses were also performed as a preliminary screening test to detect the presence of oleo-resinous material, before GC-MS analysis.

1. RESULTS AND DISCUSSION

Analytical procedures, developed and tested on standards, were applied to analyze samples collected from old masters paintings that where in restoration by the Istituto Centrale per il Restauro (ICR) of Rome, Italy. In particular, the "*Madonna con Bambino e Angeli*" by Antonello da Messina (XV century) was considered (Figure 1a).

Firstly, a micro-FT-IR was used because it proves to be a powerful tool to obtain significant and non-destructive analyses on different materials, both organic (such as binders, fibers, polymers) and inorganic (pigments and fillers for instance) [25–28]. This technique was employed as a preliminary analytical technique to detect the presence of resinous materials in a sample collected from the painting of artistic interest "*Madonna con Bambino e Angeli*", by Antonello da Messina.

FT-IR spectra of the most commonly used, in fine arts, drying oils (e.g., linseed oil) and terpenic resins (e.g., Venice turpentine, Manila copal, dannar, mastic) were collected, in order to identify diagnostic bands.

Spectroscopic analyses were performed on standards of fresh and aged resins. FT-IR spectra of natural terpenic resins are characterized by typical absorption bands, according to their molecular functional groups. Typical spectra of natural resins show a broad band in the 3500 cm^{-1}region due to the stretching of OH groups, absorption of C = O and C-O groups occur respectively at 1715–1695 cm^{-1} (strong) and 1240 cm^{-1}(weak). Methylic and methylenic groups give two sharp, strong absorption in the ranges 2960–2930 cm^{-1} and 2875–2865 cm^{-1} (C-H stretching) while 1467–1448 cm^{-1} and 1387–1382 cm^{-1} bands are due to the C-H bending [29].

Figure 1. a: " ***Madonna con Bambino e Angeli*** **" by Antonello da Messina (XV century; Uffizi Gallery, Florence). b**: layer from which the sample was taken.

The aged linseed oil, indeed, shows the following characterizing bands: 3400, 2930, 2855, 1780, 1735, 1713, 1459, 1418 cm^{-1} (strong), 1245, 1178 cm^{-1} (medium) and 1097, 980, 725 cm^{-1} (weak).

The analysis of a sample extracted by solvent action on a sheet of Japanese paper, according to the usual restorers' practice [30, 31], is an interesting example of the micro FT-IR potentialities. The analyses were carried out directly on a single fiber of the paper crashed on the diamond cell. The IR beam was focused first on a sample spot, then on a clean fiber spot. The FT-IR spectrum of the varnish was obtained subtracting the spectrum of the blank fiber from the spectrum containing both the varnish and the cellulose absorption bands.

The obtained spectrum is characterized by the following absorption bands: 3400 cm^{-1} (broad) due to the OH stretching; C = O stretching occurs at 1780, and 1730 (as shoulders) due to the presence of drying oils and 1707 cm^{-1} (strong) can be ascribed to resins. In the finger-print region the following bands occur: 1460 and 1384 cm^{-1} (medium), 1318, 1186, 1135 cm^{-1} (weak).

In the case of the painting by Antonello da Messina, it was possible, analyzing a single fiber from the Japanese paper employed to extract by solvent action the varnish, to get meaningful information on the oleo-resinous varnish extracted from the painting and on the cellulose of the fiber at the same time (Figure 2).

It is however possible to identify the chemical class of organic materials by the diagnostic bands. Here the presence of drying oils and terpenic resins was verified.

Micro-FT-IR analyses did not permit, however, to identify directly the kind of resin or of drying oil employed in a paintintheg, but allows us to address the subsequent analyses in order to optimize the employment of the micro-samples, avoiding losses of scarce and precious materials.

Figure 3 shows the Total Ion Chromatogram (TIC) of a sample taken from the painting by Antonello da Messina. The sample was collected from a layer (Additional file 1: Figure S1) corresponding to the binding medium (Figure 1b).

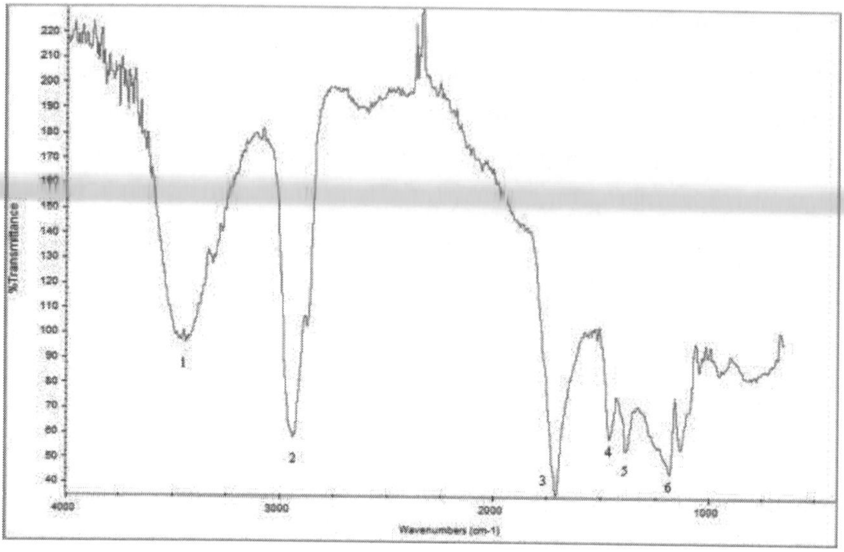

Figure 2. Micro-FTIR spectrum of the varnish from the painting, obtained by spectral subtraction of the fiber of Japanese paper from the sample. Signals: 1: OH stretching; 2: C-H bending of methylic and/or methylenic groups; 3 and 6: C = O and C-O stretching; 4 and 5: C-H bending of methylic and/or methylenic groups.

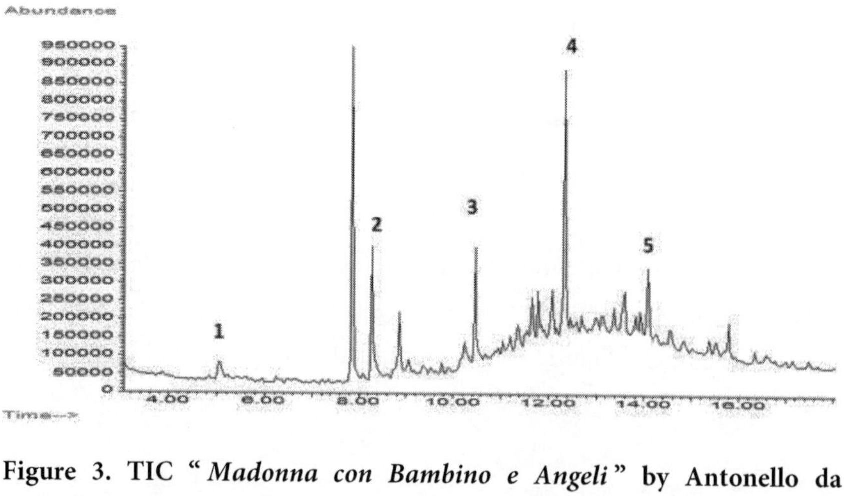

Figure 3. TIC "*Madonna con Bambino e Angeli*" by Antonello da Messina. 1: azelaic acid; 2: palmitic acid; 3: stearic acid; 4: dehydroabietic acid; 5: 7-oxo-dehydroabietic acid. For experimental condition: see the text.

The study and the characterization of binding media represent important aspects for painting characterization. In oil painting the binding medium used by artists is a drying oil which allows effective color application and helps in providing optic properties to pigments in drying place. Linseed oil, walnut oil and poppyseed oil are traditionally used in paintings. Different characteristic can be considered for identifying a binding medium. Therefore, one or more characteristic components of the oil used, should be detected. For example, a lipidic fraction may be characterized according to certain mono- or di-carboxylic fatty acids, or by using the ratio between azelaic acid and palmitic acid, or the one between palmitic acid and stearic acid. During ageing, drying oils show the production and progressive increase of dicarboxylic acids, while saturated fatty acids are so stable that they are not involved in the oil polymerization process. The palmitic and stearic acid concentrations change due to the evaporation over time but their ratio remains more or less stable. Several authors have evaluated the efficiencies of the ratio between palmitic acid and stearic acid (P/S) as an identifying parameter of linseed, walnut and poppyseed oil. They found P/S values included in the range 1.4-1.9 for linseed oil [20, 32–34]; 2.4-2.9 for walnut oil [20, 32–34]; 2.9-3.7 for poppyseed oil [34–38]. Starting from these considerations, over the chromatographic profile above reported, we found presence of both saturated acids (palmitic acid and stearic acid) and the decomposition products of the oil painting film (azelaic acid). In particular, the P/S ratio (1.3) suggests the employment of linseed oil in this painting.

Actually, also palmitic acid/azelaic acid (P/A) and palmitic acid/oleic acid (P/O) ratios are important *markers* of the ageing process but it will be not discussed in this paper.

A similar characterization was performed on the natural resin added to oil paintings for increasing flexibility in oil binders. A previous study [39] reported the detection of some characteristic molecules considered *markers* for natural resins. In particular, among the diterpenic resins, dehydroabietic acid and 7-oxo-dehydroabietic acid are characteristic for Venice turpentine. The mass spectra along with the molecular structures of these compounds are reported in Figure 4. The mass spectra of methyl esters of diterpenic abietic acids are characterized by a typical fragmentation pattern, which involves: a) the loss of the ester group; b) the expulsion of a methyl group (often in the C20 position); c) expulsion of water or methanol from the hydroxyl moieties. In Table 2 characteristic fragment ions are listed for dehydroabietic acid methyl ester and 7-oxo-dehydroabietic acid methyl ester. In Figure 5 the fragmentation patterns for these two compounds are illustrated in more details.

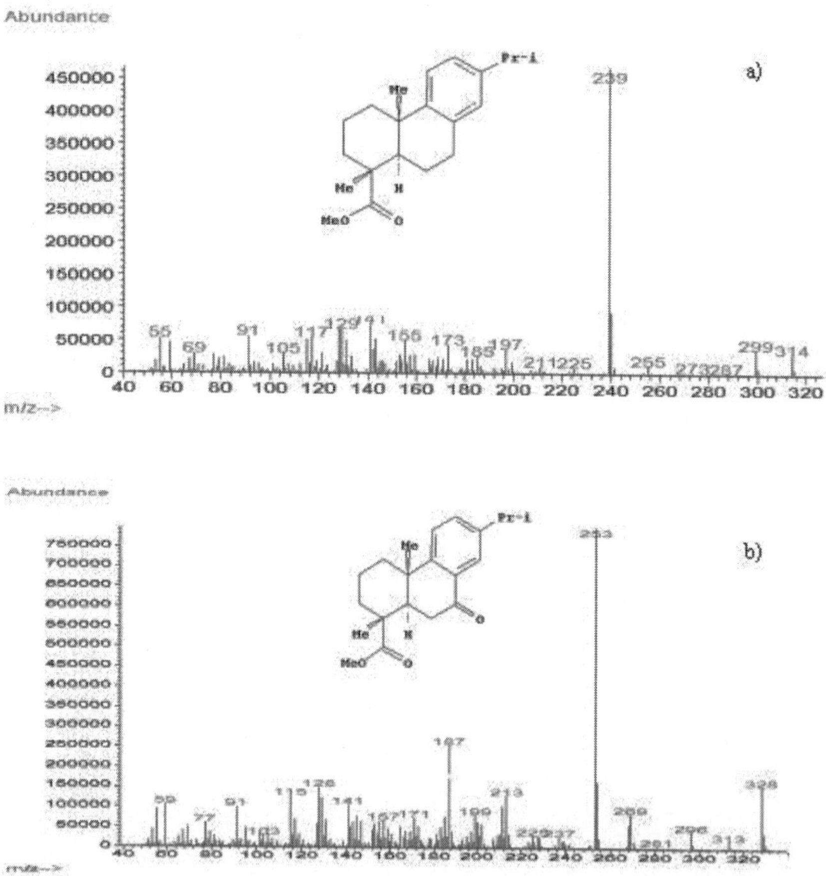

Figure 4. Mass spectra and molecular structures of dehydroabietic acid methyl ester (a) and 7-oxo-dehydroabietic acid methyl ester (b).

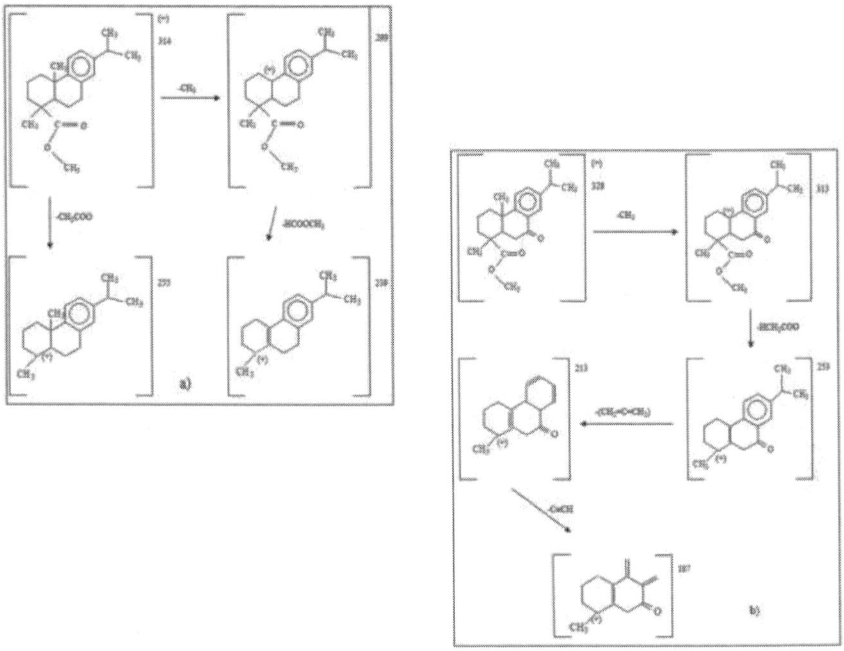

Figure 5. Fragmentation pattern of dehydroabietic acid (a) and 7-oxo-hydroxydehydroabietic acid (b).

Figure 6 show the TICs of raw (a) and aged (b) Venice turpentine, respectively: several compounds were identified in order to characterize this resin. Among them it was important to point out the species that survive to ageing processes, in order to find out suitable markers that allow us to recognize Venice turpentine in a sample. Thus, we analyzed artificially aged Venice turpentine. The artificial ageing process was performed by exposing specimens on glass of Venice turpentine, previously dissolved in methylene chloride, to the action of UV radiation for 22 days and then leaving the resin to ultimate the ageing process for 18 years in the dark [40–43].

It is interesting to notice that the composition of Venice turpentine changes significantly with ageing. The most abundant compound found in fresh resin is abietic acid because Venice turpentine is a valuable material used only in fine arts. Therefore larch resin, from which Venice turpentine is obtained, undergoes soft refining thermal treatments. Venice turpentine turns out to be softer and more malleable than other resins and retains the intense characteristic odor of the original resin.

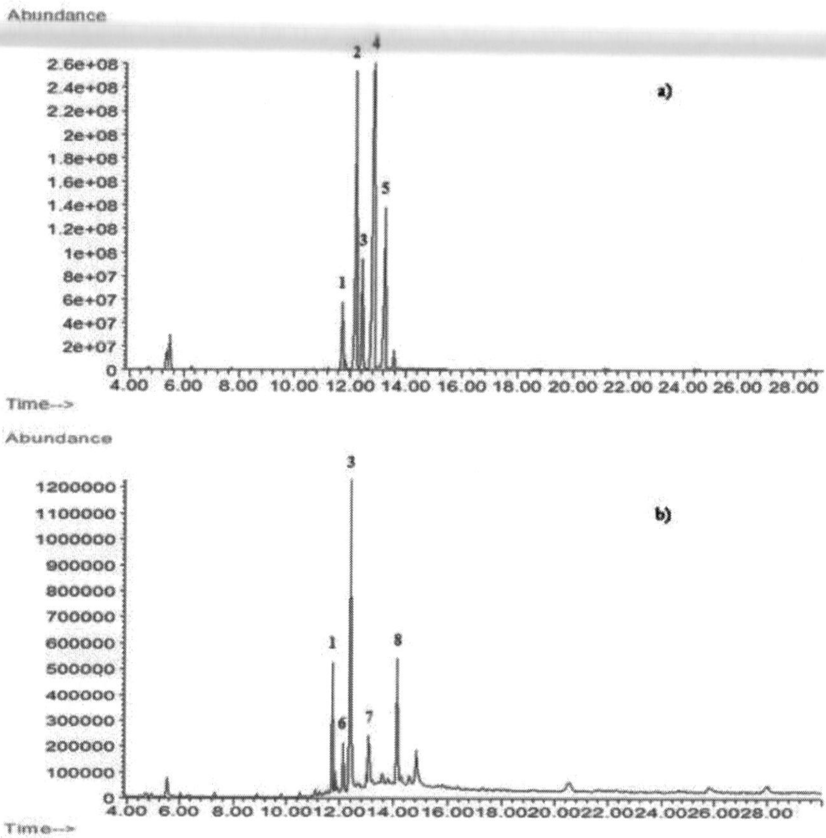

Figure 6. Partial TIC of fresh (a) and aged (b) Venice turpentine. For experimental condition: see the text. Peak number: 1: sandaracopimaric acid (Molecular Weight, MW, 316); 2: palustric acid (MW 316); 3: dehydroabietic acid (MW 314); 4: abietic acid (MW 314); 5: neoabietic acid (MW 316); 6: isopimaric acid (MW 316); 7: 6-dehydroabietic acid (MW 309); 8: 7-oxodehydroabietic acid (MW 328).

Still, in aged resin we do not find trace of abietic acid. This is another proof of the tendency to disappear of this compound with ageing. In fresh Venice turpentine dehydroabietic acid is present in small amounts, while 7-oxo-dehydroabietic acid is absent. With ageing the situation changes and dehydroabietic acid becomes the main constituent, followed by 7-oxo-dehydroabietic acid. Another significant compound which survived ageing, turned out to be sandaracopimaric acid. By the comparison of the chromatogram of fresh and aged resin it is possible to identify dehydroabietic acid, 7-oxo-dehydroabietic acid as markers of Venice turpentine.

Finally, some authors report on characteristic compounds of Venice turpentine, as epimanool, larixol and larixyl acetate [15, 20]. Actually, we did not find traces of these compounds neither in fresh samples of Venice turpentine, nor in aged ones. So we could not use these compounds as ulterior markers of the Venice turpentine.

In the painting investigated in this work, the TIC chromatogram reported in Figure 3, shows the presence of the markers of the diterpenic resin of the *pinacea* family, Venice turpentine: methyl dehydroabietate and methyl 7-oxodehydroabietate. It is very interesting to notice that the sample was taken from an original painting layer since, when the sample was taken, the superficial varnish layer had already been removed.

So we could state that the artist originally used a mixture of linseed oil and Venice turpentine (natural resin) as binding medium.

2. EXPERIMENTAL

2.1. Materials

The reagent used were: dichloro methane, methyl alcohol (RS for HPLC), diethyl ether, KOH and HCl (Carlo Erba, Milano, Italy). Dammar resin and elemi (G-9877 and G-0263, respectively) were supplied by Sigma (Milan, Italy). Venice turpentine (art. 3230) was purchased from Zecchi (Florence, Italy). Both sandarac resin and Manila Copal were supplied by Morrone (Rome, Italy). A sample of resin exudates was collected from Larix deciduas as fresh raw material. Venice turpentine, dammar, elemi and Manila Copal were also used for specimens on glass. Resins spread on glass supports were artificially aged by the action of UV radiation. A 150 W quartz sheathed mercury vapor lamp "H" (Helios Italquartz s.r.l., Milano, Italy), maximum emission at 320 nm and 370 nm, was employed to irradiate natural resins for 22 days at the distance of 50 cm in a weathering case type UY 600 (Angelantoni Climatic Systems, Massa Martana, Perugia, Italy) at 25°C and 80% Relative Humidity (RH); then the above mentioned resins were allowed to ultimate their ageing process for 18 years in the dark.

A sample of the paint layer from an old master oil paintings on canvas, was taken. It was collected from the painting "Madonna con Bambino e Angeli" by Antonello da Messina (ca. 1470), housed in "Uffizi" Firenze (Italy), and analyzed (Figure 1a). The sample was collected from the uppermost layer after that the superficial varnish layer has already been removed. This means that the sample was taken from an original painting layer (Figure 1b).

2.2. Sample preparation for GC-MS analyses of standard resins

The procedure followed for standard resins samples preparation involved the sample dissolution in dichloro methane and subsequent derivatization by diazomethane [44].

1 mL of dichloro methane was added to a known quantity (1 mg) of sample. After the complete dissolution of the sample, diazomethane in diethyl ether solution was added till the solution turned deep yellow for the excess of diazomethane. The solution was then evaporated to 150 μL to eliminate the diazomethane in excess under a mild flow of N_2.

2.3. Sample preparation for GC-MS analyses of material collected from paintings

The samples collected from the old masters paintings could contain both oils and resins, so a double step derivatization method was followed. This procedure involved the alkaline hydrolysis of the triglycerides contained in drying oils and subsequent derivatization of the fatty acids obtained, these were then turned into methyl esters by trans-esterification reaction [35] and extracted with dichloro methane. The methyl esters were extracted using n-hexane (0.5 mL × 3 times). The second derivatization step was performed by the addition of diazomethane in diethyl ether solution to the collected extracts to improve the derivatization for carboxylic functions of resin components.

100 μL of dichloro methane and 100 μL of KOH 2 M methanolic solution were added to the sample (< 0.1 mg). The solution was stirred for about 2 minutes and left to rest for 15 minutes at 60°C. Then 500 μL of a solution of HCl_{conc} and CH_3OH (1:1) were added. The esters were extracted three times using 100 μL of dichloro methane. Few drops of diazomethane in ethyl ether solution was added to the collected extracts, according to the previously described procedure. The solution was evaporated to 150 μL under a mild flow of N_2, and were analyzed by GC-MS.

2.4. GC-MS apparatus

GC-MS analyses were performed on a HP-5890 Series II gas chromatograph coupled to a HP-5972 mass selective detector. A fused-silica capillary column with chemically bonded phase (SE-54, 5% phenyl-95%dimethylpolysiloxane) was prepared in our laboratory [45, 46] with the following characteristics: 30 m × 250 μm i.d., N (theoretical plate number) 115,000 for *n*-dodecane at 90°C; K' (capacity factor) 6.7; d_f (film thickness) 0.24 μm; u_{opt} (optimum linear velocity of carrier gas hydrogen) 39.0 cm s^{-1}, and UTE% (utilization of theoretical efficiency) 91%. The chromatographic conditions for GC-MS analysis were: injector temperature 300°C; transfer line temperature 280°C; initial oven temperature 120°C; isothermal for 1 min; 30°C min^{-1} up to 200°C and isothermal for 2 min; then 5°C min^{-1} up to the final temperature of 240°C. The carrier gas was helium at a flow-rate of 1.3 mL min^{-1}. The GC-MS analysis was performed in the SCAN mode (mass range 50–550 amu at 70 eV). The terpenic compound identification was obtained by the comparison of mass spectra included in the library NBS 75 K.1 of the mass selective detector UP-5972. Spectra not included in above mentioned library were compared with mass spectra available in literature [20, 47–52].

2.5. Micro-FT-IR apparatus and analyses

Micro-FT-IR analyses were made on a Nicolet (Madison, Wisconsin ,USA) 510 P FT-IR spectrometer coupled to a Nicolet IR microscope, mod. NicPlan. A Mercury-Cadmium-Telluride (MCT) detector, cryogenically cooled, was used to examine the region from 4000–650 cm^{-1}. Each recorded spectrum is the sum of sixty four scans collected at a resolution of 4 cm^{-1}.

The samples, crushed in a sample compression cell with diamond windows (μSample PlanTM, mod. 0042–444, Spectra-Tech Inc., Stamford, Connecticut, USA), were analyzed in transmission mode. The diamond cell, containing the crushed sample, was mounted on the microscope stage at the focal point of the IR beam condenser for spectroscopic analysis.

3. CONCLUSIONS

The study of molecular changes due to ageing processes is particularly important, because these transformations cause physical changes such as variation in solubility. This aspect of the problem has to be taken into account when treatments to remove varnishes are applied.

Micro FT-IR technique was sometimes employed as a preliminary screening test to detect the presence of organic materials, drying oils and resins in particular, in microsamples collected from the paintings. The information obtained through IR spectra allowed us to focus the chromatographic analysis on these kinds of substances, thus getting the maximum information on the painting, using the minimum amount of sample. This is a very important point as the first aspect to consider when we approach the study of a work of art is its preservation.

As natural resins could be found in binding media as trace components, it is important to test procedures to analyze resins efficiently. Moreover it could be useful to develop analytical methods to determine, at the same time, the presence of oils and resins in micro-samples drawn from paintings.

GC-MS analyses of the standards of oils, both fresh and artificially aged, allowed us to verify the importance of the ratio between palmitic acid and stearic acid as *marker* for the identification of the particular drying oil used.

The analytical procedure developed and the results of the study carried out on standard materials allowed us to identify the drying oil and, eventually, the varnish used in the painting of artistic interest (XV century).

All the considerations reported in this paper, should help the restorers, not only from the point of view of art history, but also from that of the chemical composition of the originally employed material in order to formulate a correct diagnosis of the state of conservation of a work of art.

AUTHORS' INFORMATION

[a]Full Professor, Faculty of Agriculture, Università del Molise, via De Sanctis, Campobasso, I-86100 Italy.

[b]Researcher, DIPIA, INAIL ex-ISPESL, via Urbana 167, Rome, I-00184 Italy.

ACKNOWLEDGEMENTS

The authors wish to thank dr. F. Talarico of the Istituto Centrale per il Restauro (ICR) of Rome for giving us some real samples, standard resins and the photos.

AUTHORS' CONTRIBUTIONS

MV coordinated the study. PA and MV set up the analytical procedure using GC-MS. PA processed data. MV provided the comparison with other literature. PA edited the text and prepared the final draft of the paper. All the authors have read and approved the final manuscript.

REFERENCES

1. Casoli A, Palla G: **Proteinaceous binding media characterization and age estimation of mediaeval wall paintings.** *Sci Tech Cult Herit* 1994, **3**:177–183.

2. Chiantore O, Riedo C, Scalarone D: **Gas chromatography–mass spectrometric analysis of products from on-line pyrolysis/silylation of plant gums used as binding media.** *Inter J Mass Spectrom* 2009, **284:**35–41.

3. Sciutto G, Dolci LS, Buragina A, Prati S, Guardigli M, Mazzeo R, Roda A: **Development of a multiplexed chemiluminescent immunochemical imaging technique for the simultaneous localization of different proteins in painting micro cross-sections.** *Anal Bioanal Chem* 2001, **399:**2889–2897.

4. Sarmiento A, Pérez-Alonso M, Olivares M, Castro K, Martínez-Arkarazo I, Fernández LA, Madariaga JM: **Classification and identification of organic binding media in artworks by means of Fourier transform infrared spectroscopy and principal component analysis.** *Anal Bioanal Chem* 2011, **399:**3601–3611.

5. Grzywacz CM: **Identification of proteinaceous binding media in paintings by amino acid analysis using 9-fluorenylmethyl chloroformate derivatization and reversed-phase high-performance liquid chromatography.** *J Chromatogr A* 1994, **676:**177–183.

6. Castro RM, Carbò MTD, Martìnez VP, Adelantado JVG, Reig FB: **Study of binding media in works of art by gas chromatographic analysis of amino acids and fatty acids derivatized with ethyl chloroformate.** *J Chromatogr A* 1997, **778:**373–381.

7. Peris-Vicente J, Gimeno Adelantado JV, Carbó MTD, Castro RM, Reig FB: **Characterization of proteinaceous glues in old paintings by separation of the o-phtalaldehyde derivatives of their amino acids by**

liquid chromatography with fluorescence detection. *Talanta* 2006, **68:**1648–165.

8. Rezić I, Krstić D, Bokić L: **A note on the determination of the binder composition on an historic painted textile.** *Stud Conser* 2006, **51:**223 228.

9. Checa-Moreno R, Manzano E, Mirón G, Capitan-Vallvey LF: **Comparison between traditional strategies and classification technique (SIMCA) in the identification of old proteinaceous binders.** *Talanta* 2008, **75:**697–704.

10. Fremout W, Sanyova J, Saverwyns S, Vandenabeele P, Moens L: **Identification of protein binders in works of art by high-performance liquid chromatography-diode array detector analysis of their tryptic digests.** *Anal Bioanal Chem* 2009, **393:**1991–1999.

11. Baggio L, Colalucci G, Bartoletti D: *Il restauro degli affreschi.* De Luca Editori d'Arte, Rome; 1999. (in Italian)

12. Mills JS, White R: **Natural resins of art and archaeology. Their sources, chemistry and identification.** *Stud Conserv* 1977, **22:**12–31.

13. Gimeno-Adelantado JV, Mateo-Castro R, Doménech-Carbó MT, Bosch-Reig F, Doménech-Carbó A, De la Cruz-Cañizares J, Casas-Catalan MJ: **Analytical study of proteinaceous binding media in works of art by gas chromatography using alkyl chloroformates as derivatising agents.** *Talanta* 2002, **56:**71–77.

14. De la Cruz-Cañizares J, Doménech-Carbó MT, GimenoAdelantado JV, Mateo-Castro R, Bosch-Reig F: **Suppression of pigment interference in the gas chromatographic analysis of proteinaceous binding media in paintings with EDTA.** *J Chromatogr A* 2004,**1025:**277–285.

15. Mills JS, White R: *The organic chemistry of museum objects.* Buttersworth, London; 1994.

16. Peris-Vicente J, Simó-Alfonso E, Gimeno Adelantado JV, Doménech Carbó MT: **Direct infusion mass spectrometry as a fingerprint of protein-binding media used in works of art.** *Rapid Commun Mass Spectrom* 2005, **19:**3463–3467.

17. Quezada-Gallo J-A, Debeaufort F, Voilley A: **Mechanism of aroma transfer through edible and plastic packagings: Are they complementary to solve the problem of aroma transfer?** *ACS Symp Ser* 2000, **753:**125–140.

18. Thomson R: *Conservation of leather and related materials.* , Buttherworth, London; 2005. pp. 88–91. ISBN: 0750648813

19. Prati S, Sciutto G, Mazzeo R, Torri C, Fabbri D: **Application of ATR-far-infrared spectroscopy to the analysis of natural resins.** *Anal Bioanal Chem* 2011, **399:**3081–3091.

20. Van den Berg K, Boon JJ, Pastorova I, Spetter LFM: **Mass spectrometric methodology for the analysis of highly oxidized diterpenoid acids in Old Master paintings.** *J Mass Spectrom* 2000, **35:**512–533.

21. Papageorgiou VP, Bakola-Christianopoulou MN, Apazidou KK, Psarros EE: **Gas chromatographic-mass spectroscopic analysis of the acidic triterpenic fraction of mastic gum.** *J Chromatogr A* 1997, **769:**263–273.

22. Chiavari G, Fabbri D, Prati S: **Characterisation of natural resins by pyrolysis – Silylation.** *Chromatographia* 2002, **55:**611–616.

23. Mizzonia F, Cesaro SN: **Study of the organic residue from a 2600-year old Etruscan plumpekanne.** *Spectrochim Acta A* 2007,**68:**377–381.

24. Johnstone RAW, Rose ME: *Mass spectrometry for chemists and biochemists.* Cambridge University Press, Cambridge; 1996.

25. van Zelst LV, von Endt DW, Baker MT: **Non-destructive and micro-sample FT-IR spectrometric analysis of organic materials in art objects.** *Proc 2nd Inter Conf Non-Destruc Test, Microanal Meth Environ Eval Study Cons Works Art* 1988, **I:**1–17.

26. Salvadó N, Butí S, Tobin MJ, Pantos E, Prag AJNW, Pradell T: **Advantages of the use of SR-FT-IR microspectroscopy: Applications to cultural heritage.** *Anal Chem* 2005, **77:**3444–3451.

27. Echard J-P, Cotte M, Dooryhee E, Bertrand L: **Insights into the varnishes of historical musical instruments using synchrotron micro-analytical methods.** *Appl Phys A* 2008, **92:**77–81.

28. Zeng QG, Zhang GX, Leung CW, Zuo J: **Studies of wall painting fragments from Kaiping Diaolou by SEM/EDX, micro Raman and FT-IR spectroscopy.** *Microchem J* 2010, **96:**330–336.

29. Derrick M: **Fourier Transform Infrared spectral analysis of natural resins used in furniture finishes.** *JAIC* 1989, **28:**43–56.

30. Ballofet N, Hille J: *Preservation and conservation for libraries and archives.* American Library Association, Chicago; 2005. ISBN 0–8389–0879–9

31. The E: **Lingle Craig Preservation Laboratory Repair and Enclosure Treatment Manual.** ,:. available at http://www.indiana.edu/~libpres/manual/ (access on May 15th, 2012)

32. Mills JS, White R: **Analyses of paint media.** *Nat Gal Tech Bull* 1980, **4:**65–67. ISSN: 0140–7430.

33. Mills JS, White R: *Application of science in examination of works of art.* Museum of Fine Arts Boston, Boston; 1985. p. 29. ISBN: 0878462554

34. Nachtigal M, Simunkova E, Zelinger J: **The influence of cobalt blue pigments on the drying of linseed oil.** *JSTOR* 1985, **30:**161–166.

35. Cartoni GP, Russo MV, Spinelli F, Talarico F: **Characterization of fatty acids in drying oils used in paintings on canvas by GC and GC-MS analysis.** *Ann Chim (Rome)* 2001, **91:**719–726.

36. Rioux JP: **Note sur l'analyse de quelques enduits provenant de peintures francais des XVII et XVIII siècles.** *Ann Musées France*1973, **3:**35–43.

37. Delbourgo S, Rioux JP: **Manière et matière des impressionistes. Contribution à l'étude de la matière picturale.** *Ann Musées France*1974, **4:**34–42.

38. Mills JS, White R: *Conservation and restoration of pictorial art.* , Butterworths, London; 1976. p. 72. ISBN: 0408707127

39. Cartoni G, Russo MV, Spinelli F, Talarico F: **GC-MS characterization and identification of natural terpenic resins employed in works of art.** *Ann Chim (Rome)* 2004, **94:**767–782.

40. Colombini MP, Modugno F, Menicagli E, Fuoco R, Giacomelli A: **GC-MS characterization of proteinaceous and lipid binders in UV aged polychrome artifacts.** *Microchem J* 2000, **67:**291–300.

41. Marengo E, Liparota MC, Robotti E, Bobba M, Gennaro MC: **Monitoring of pigmented surfaces in accelerated ageing process by ATR-FT-IR spectroscopy and multivariate control charts.** *Talanta* 2005, **66:**1158–1167.

42. Manzano E, Navas N, Checa-Moreno R, Rodriguez-Simón L, Capitán-Vallvey LF: **Preliminary study of UV ageing process of proteinaceous paint binder by FT-IR and principal component analysis.** *Talanta* 2009, **77:**1724–1731.

43. Boeva R, Radeva G, Hinkov P, Hinkov E: **Thermal ageing of different kinds of fiber materials: a kinetic study.** *J Univ Chem Technol Met* 2012, **47:**37–42. available at http://www.uctm.edu/journal/j2012-1/3_Boeva_37-42.pdf

44. Cartoni G, Russo MV, Spinelli F, Talarico F: **Characterization of fresh and aged terpenic resins by micro-FTIR and GC-MS analyses of varnishes in XVI-XVII centuries paintings.** *Ann Chim (Rome)* 2003, **93:**849–861.

45. Cartoni GP, Goretti G, Neri B, Russo MV: **Evaluation of small diameter capillary columns for gas chromatography.** *J Chromatogr* 1989, **475:**145–151.

46. Russo MV, Goretti G, Soriero A: **Preparation and application of fused-silica capillary microcolumns (25–50 µm I. D.) in gas chromatography.** *Ann Chim (Rome)* 1996, **86:**115–124.

47. Chang T-L, Mead TE, Zinkel DF: **Mass spectra of diterpene resin acid methyl esters.** *J Am Oil Chem Soc* 1971, **48:**455–461.

48. Shiojima K, Arai Y, Masuda K, Takase Y, Ageta T, Ageta H: **Mass spectra of pentacyclic triterpenoids.** *Chem Pharm Bull* 1992,**407:**1683–1690.

49. Van der Doelen GA: *Molecular studies of fresh and aged triterpenoid varnishes. Ph.D. thesis.* University of Amsterdam, ; 1999. ISBN: 90-801704-3-7

50. van der Doelen GA, van den Berg KJ, Boon JJ, Shibayama N, de la Rie ER, Genuit WJL: **Analysis of fresh triterpenoid resins and aged triterpenoid varnishes by high-performance liquid chromatography–atmospheric pressure chemical ionisation (tandem) mass spectrometry.** *J Chromatogr A* 1998, **809:**21–37.

51. Dietemann P, Higgitt C, Kälin M, Edelmann MJ, Knochenmuss R, Zenobi R: **Aging and yellowing of triterpenoid resin varnishes - Influence of aging conditions and resin composition.** *J Cult Herit* 2009, **10:**30–40.

52. Yang M, Wang X, Guan S, Xia J, Sun J, Guo H, Guo D: **Analysis of triterpenoids inGanoderma lucidumusing Liquid Chromatography coupled with Electrospray Ionization Mass Spectrometry.** *J Am Soc Mass Spectrom* 2007, **18:**927–939.

CHAPTER 6

Analysis of hydrazine in smokeless tobacco products by gas chromatography–mass spectrometry

Kevin McAdam[1], Harriet Kimpton[1], Sofia Essen[1], Peter Davis[1], Carl Vas[1], Christopher Wright[1], Andrew Porter[2] and Brad Rodu[3]

[1]*Group Research & Development, British American Tobacco, Regents Park Road, Southampton SO15 8TL, UK.*

[2]*3810 StAntoine W, Montreal, QC H4C 1B4, Canada.*

[3]*Room 208, 505 South Hancock Street, Louisville, KY 40202, USA*

ABSTRACT

Background

Due to the lower health risks associated with the use of certain categories of smokeless tobacco products (STPs) such as Swedish snus, there is interest in the comparative levels of toxic chemical constituents in different types of STPs. A method has been developed and validated for the analysis of hydrazine in STPs. Seventy four commercial STPs from the US and Sweden, representing 80-90% of the 2010 market share for all the major STP categories in these two countries, as well as three reference STPs, were analysed for hydrazine.

Results

Aqueous extracts of the STPs were treated with excess pentafluorobenzaldehyde (PFB), which reacted with hydrazine in solution to form decafluorobenzaldehyde azine (DFBA). DFBA was partitioned into hexane and then quantified by gas chromatography–mass spectrometry (GC–MS). The method was validated using five different types of STP, was linear in the range 8–170 ng/mL, and had limits of quantification (LOQ) from 26–53 ng of hydrazine per g of STP (as sold). The method was applied to the analysis of 74 contemporary STPs commercially available in the United States and Sweden, none of which were found to contain hydrazine above the LOQ or LOD. Trace levels of compounds showing chromatographic and mass spectral features consistent with hydrazine were identified at very low levels (sub-limit of detection, <10 ng/g) in the chromatograms of less than half of the 74 STPs examined; in contrast, for 40 of the STPs no evidence for the presence of hydrazine was observed. Where present, the levels of compounds consistent with hydrazine were estimated to be at least an order of magnitude lower than the only previous study to have quantified hydrazine in tobacco.

Conclusions

Our results show that hydrazine is not a prevalent constituent of STPs, and when present is not quantifiable using currently available analytical methodology.

KEYWORDS

Hydrazine Smokeless tobacco products Snus Snuff

1. BACKGROUND

Smokeless Tobacco Products (STPs) have been collectively designated as Group 1 carcinogens i.e. carcinogenic to humans [1], but there is considerable evidence that health risks differ between STP categories, with certain product styles such as Swedish snus having lower health risks associated with their use [2]. As a result, there is substantial interest in the comparative levels of toxic chemical constituents of the different types of STPs. In the US, the Food and Drug Administration (FDA) has assembled a list ("The Established List") of 93 "hazardous or potentially hazardous constituents" (HPHC) of tobacco products which may have to be reported [3]. This list covers both tobacco and tobacco smoke components and includes 79 that are designated as carcinogenic as well as constituents that are respiratory toxicants, cardiovascular toxicants, reproductive toxicants or addictive. One of the HPHC carcinogens on the list is hydrazine (N_2H_4) which has been classified as a group 2B carcinogen (possibly carcinogenic to humans) by IARC [4].

Although studies of hydrazine toxicity in humans are limited, human exposure to hydrazine has resulted in severe effects on the central nervous system, liver and kidneys [4]. Hydrazine is mainly an industrial chemical, manufactured from ammonia, that can enter the environment from facilities that manufacture, process or use it. Hydrazine is unstable and degrades rapidly in most environmental media. It can dissolve in water and move though soil, but hydrazine is broken down by autoxidation and by microorganisms. A review [5] of three studies concluded that half-lives of hydrazine in soil ranged from 1 hr to 3 days with the more rapid degradation of hydrazine occurring in soils with high levels of microorganisms and organic material.

The occurrence of free hydrazines in nature is rare. Naturally occurring hydrazine and hydrazone derivatives such as agaritine (β-N-[γ-L(+)-glutamyl]-4-hydroxymethyl phenylhydrazine) and gyromitrin (acetaldehyde methylformylhydrazone) have been reported in mushrooms [6]. Gyromitrin breaks down during cooking to release methylhydrazine but the latter is not found in a free state in the mushroom. Hydrazine is produced as an intermediate during biological nitrogen fixation by the molybdenum- and vanadium-based nitrogenase enzymes in *Azotobacter*[7]. In the case of the more abundant molybdenum-based nitrogenase the hydrazine is bound to the enzyme and is not released in a free state. However for vanadium- based nitrogenase small but significant amounts of free hydrazine are generated

[8]. There is therefore the potential for hydrazine to be found in plant materials that are associated with nitrogenase containing bacteria.

The only occurrence of free hydrazine in plant material was that reported in tobacco by Liu et al. [9]. They found small amounts of hydrazine in tobacco from a commercial cigarette (30.0 ng/cigarette) and in the tobacco of four experimental cigarettes. Two of the experimental cigarettes were made with Burley tobaccos, one treated with the plant sucker growth inhibitor maleic hydrazide (MH) (51.2 ng hydrazine/cigarette) and one untreated (22.2 ng hydrazine/cigarette); the other two cigarettes were made with flue-cured tobacco, one treated (12.1 ng hydrazine/cigarette) and one untreated (13.8 ng hydrazine/cigarette). Liu et al. [9] also determined hydrazine in the mainstream smoke of these 5 cigarettes (range 23.5–42.8 ng/cigarette). The hydrazine concentrations in tobacco and tobacco smoke obtained in the original Liu et al. study of over 40 years ago [9], have been frequently reproduced in review articles [10-16]. No other study of hydrazine in tobacco has been reported, although several other studies have failed to detect hydrazine in tobacco smoke [17-19]. Using the same methodology as Liu et al. for hydrazine, Schmeltz et al. [13] found the hydrazine derivative, 1,1-dimethylhydrazine, in several samples of tobacco including US chewing tobacco (97.7 ng/g) and snuff (96.7 ng/g), four commercial US cigarette blends (60.2 ± 5.7 ng/g) and Bright tobacco (147 ng/g). No 1,1-dimethylhydrazine was found in a sample of Burley tobacco.

We are currently conducting a comprehensive survey of toxicants in an extensive and varied set of contemporary STPs from the United States and Sweden. There have been no further published studies of hydrazine in tobacco since the report by Liu et al. in 1974, and no studies of hydrazine in STPs have ever been reported. The aims of the present study were therefore to develop and validate a method for the analysis of hydrazine in STPs, and to survey major STPs from the USA and Sweden for their hydrazine content.

Various methods have been reported for the detection of trace levels of hydrazine in substrates such as sludge, human plasma, environmental water and drug samples. These have included chemical derivatization with reagents such as benzaldehyde, 4-hydroxy benzaldehyde, 2-

hydroxynaphthaldehyde, 2,4-dinitrochlorobenzene and acetone or acetone-d6, coupled with HPLC and/or spectrophotometric detection [20-23], solid-phase spectrophotometry [24], and GC–MS [25]. Indirect methods have also been developed, such as oxidation of hydrazine by excess iodate [26] or bromine [27], and analysis of the unreacted oxidant. The detection limits for these methods vary from 1 ng/ml [21] to 100 ng/g [25]. The analytical method used in the original study in tobacco by Liu et al. [9] was based on the reaction of residues of hydrazine in tobacco product extract with PFB to form DFBA, Figure 1, coupled with multiple thin-layer chromatographic steps, followed by GC separation and electron capture detection.

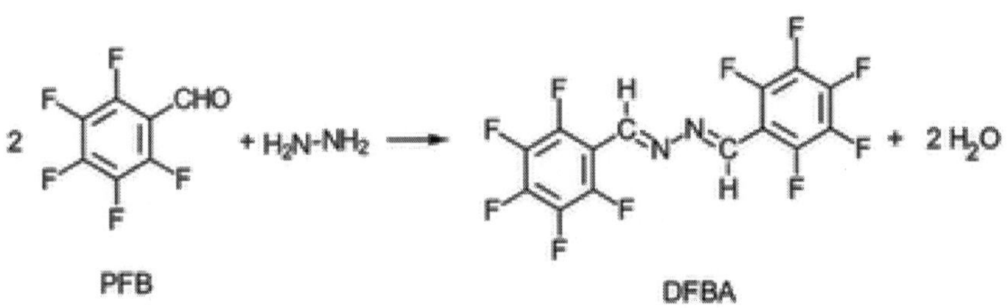

Figure 1. Reaction of pentafluorobenzaldehyde (PFB) with hydrazine to form decafluorobenzaldehyde azine (DFBA).

For the analysis of STPs, we chose to use derivatization of hydrazine with PFB, as described in the Liu et al. paper, but coupled with GC–MS to improve the sensitivity and to eliminate the need for the multiple concentration steps used in the original paper. The present method was based on the reaction of residues of hydrazine in tobacco product extract with PFB to form DFBA. The DFBA is partitioned into hexane and then quantified by GC–MS. After validation of this approach, the method was applied to the analysis of 74 contemporary STPs commercially available in the United States and Sweden. The products covered all major STP categories and the brands selected represented 90% market share of the major product styles [28].

2. RESULTS AND DISCUSSION

2.1. Validation of the analytical method

As a first step in validating the analytical approach, the identity of the peak assigned to DFBA was confirmed by visual examination of the chromatograms and mass spectra obtained for standards and tobacco samples spiked with hydrazine at 0.53 μg/g. The retention time of the GC peak assigned to DFBA was 9.9 min for all standards and types of STP. The mass spectra of the assigned peaks were almost identical for all standards and spiked tobacco samples and included ion clusters at m/z 388 (molecular ion and base peak), 194, 180, 117 and 93. Lastly, the MS software selected the spectrum of DFBA from the "Saturn library" as the closest match to that of the chromatogram peak.

To check the linearity of the method, six standard DFBA solutions, ranging from 100 to 2000 ng/mL (equivalent to 8–170 ng/mL hydrazine), were applied to GC–MS in a random order. The areas of the peaks were subjected to linear regression analysis. The analysis was highly linear across the six standards with R^2 values in excess of 0.99 (Figure 2).

In a preliminary analysis, the five control tobacco samples (Garrett, Stonewall Wintergreen, Ettan Loose, Days Work and Taylor's Pride), each representing a different type of STP, were analysed in duplicate. All of these STPs had apparent hydrazine contents below the lowest standard analysed (100 ng/mL DFBA), corresponding to concentrations of <16 ng/g.

To test the accuracy and precision of the analytical method, analyses were carried out in which the five control STPs were spiked with hydrazine at 530 ng/g, 53 ng/g and 26.5 ng/g. For each STP at each level of spiking, five repeat analyses were carried out (Table 1).

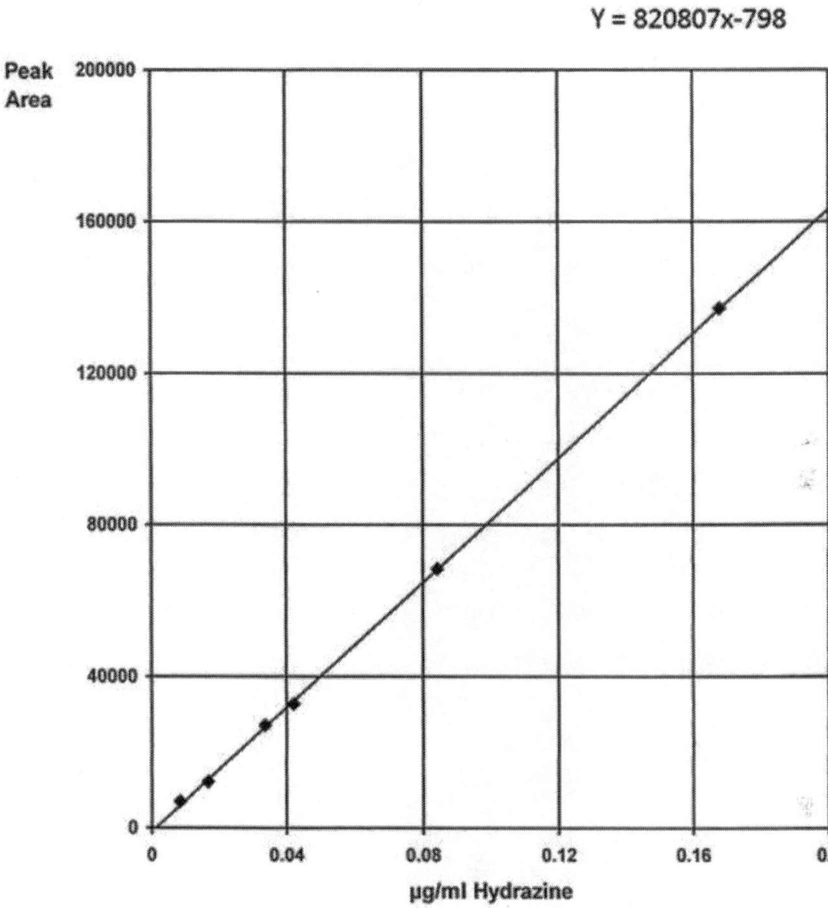

Figure 2. Linearity of the analytical method. The intercept on the *x* axis represents a hydrazine concentration of 0.0018 µg/mL.

At spike levels of 530 and 53 ng/g hydrazine, the mean recoveries from all tobacco product types were within the acceptable range [29] of 70%–110% (Table 1), indicating satisfactory accuracy. At the spike level of 26.5 ng/g hydrazine, a low recovery of 64% was observed for Stonewall Wintergreen. The hard pellet has a high mineral content that may have affected the hydrazine added, either through chemical reaction or adsorption. With the %RSD below 20%, the precision of the analytical technique was satisfactory at all three spike levels.

Table 1. Recovery and repeatability of measurements of spiked tobacco samples

Sample	Style	Hydrazine spike level (ng/g)	Percentage recovery, mean ± RSD (n = 5)	Overall percentage RSD
Garrett	Dry snuff	530	87.5 ± 5.1	
		53.0	105.2 ± 4.7	10.7
		26.5	96.7 ± 11.9	
Stonewall Wintergreen	Pellet	530	77.0 ± 5.0	
		53.0	74.1 ± 6.9	10.8
		26.5	64.0 ± 11.3	
Ettan	Loose snus	530	75.7 ± 8.3	
		53.0	96.1 ± 6.9	11.9
		26.5	83.1 ± 3.3	
Days Work	Plug	530	85.2 ± 4.1	
		53.0	95.3 ± 8.5	7.8
		26.5	92.2 ± 6.2	
Taylor's Pride	Chewing tobacco	530	90.5 ± 12.0	
		53.0	98.3 ± 3.2	9.15

Sample	Style	Hydrazine spike level (ng/g)	Percentage recovery, mean ± RSD (n = 5)	Overall percentage RSD
		26.5	99.7 ± 9.1	

The LOQ was defined as the lowest spike level for which there was an acceptable recovery (i.e. in the range 70%-110%). The LOQ was therefore 26.5 ng/g for the chewing tobacco, dry snuff, moist snuff and loose snus, and 53 ng/g for the hard pellet. The limit of detection (LOD) was estimated to be less than 10 ng/g from this validation. The linearity, LOQ, LOD, precision, and accuracy of the method are summarized in Table 2.

Table 2. Linearity and sensitivity of the analytical method for hydrazine

Validation parameter	Value
Linear range	8–170 ng/mL
Linear curve	$y = 1.03 \times 10^6 x - 1886.88$
R^2	0.9996
Accuracy across 5 types of STP (% mean ± RSD recovery)	64.0%–105.2%
Accuracy across 4 types of STP[a] (% mean ± RSD recovery)	75.7%–105.2%
Precision across 5 types of STP (%RSD of repeatability of spike recovery)	7.8%–11.9%

Validation parameter	Value
LOQ for 5 types of STP	53.0 (ng/g)
LOQ for 4 types of STP[a]	26.5 (ng/g)
LOD	<10 (ng/g)

[a]Without the pellet product, which showed poor recovery.

2.2. Survey of 74 STPs for hydrazine

Once validated, the method was used to survey the 74 contemporary STPs for hydrazine levels. As shown in Tables 3 and 4, none of the products was found to contain hydrazine at levels above the LOD (<10 ng/g). Examination of the chromatograms identified a very low level peak at the retention time of hydrazine (Figure 3A, C), and with a matching mass spectrum, for 34 of the 74 STPs analysed. The peaks were sufficiently infrequent, and below the LOD, that the possibility cannot be discounted that these peaks were merely analytical noise. However, as the peaks showed identical chromatographic and mass spectral features to hydrazine, we regard it as possible that very low levels of hydrazine were present in the samples showing these peaks. Of these STPs, the majority (22) showed the hydrazine peak in only one of the three replicates analysed, seven STPs showed the peak in two of the three replicates, and five STPs showed the peak in all three replicates. In total 51 of the 222 replicate analyses showed the presence of hydrazine, and therefore the majority of analyses showed no evidence for the presence of low levels of hydrazine (Figure 3B, D, E). Clearly, if there is interest in quantifying these potential very low levels of hydrazine, a much more sensitive analytical method would be required, with more than an order of magnitude greater sensitivity. It is unlikely that the current approach is modifiable to this extent, and alternative approaches may be required.

Table 3. Estimated hydrazine concentrations in Swedish STPs

Swedish STPs	Style	Hydrazine content (ng/g)**	No of replicates with peaks below LOD corresponding to hydrazine
Ettan	Loose snus	<LOD	1
General		<LOD	0
Goteborgs Rape		<LOD	0
Granit		<LOD	0
Grovsnus		<LOD	3
Knox		<LOD	0
Kronan		<LOD	0
LD Original		<LOD	0
T. Montecristo		<LOD	0
Skruf Strong		<LOD	1
Catch Licorice, mini	Portion snus	<LOD	0
Catch White Licorice		<LOD	2
CatchDry White Eucalyptus, mini		<LOD	1

Swedish STPs	Style	Hydrazine content (ng/g)**	No of replicates with peaks below LOD corresponding to hydrazine
Ettan		<LOD	0
General		<LOD	0
General mini		<LOD	0
General White		<LOD	0
Goteborgs Rape		<LOD	0
Granit		<LOD	1
Granit White		<LOD	0
Grovsnus		<LOD	0
Grovsnus White		<LOD	0
Gustavus Original		<LOD	1
Knox		<LOD	0
Kronan		<LOD	0
LD Original		<LOD	0
Oomph Citrus Menthol		<LOD	1

Swedish STPs	Style	Hydrazine content (ng/g)**	No of replicates with peaks below LOD corresponding to hydrazine
Romeo y Julieta Habanos		<LOD	0
Skruf Strong		<LOD	1
Tre-Ankare White		<LOD	1
1847 Original		<LOD	0
CRP1		<LOD	0

All STPs have mean hydrazine contents below the LOD (<10 ng/g STP "as sold"*).

*All hydrazine concentrations are based on the "as sold" or wet weight of STP, with no corrections for moisture.

Table 4. Estimated hydrazine concentrations in US STPs

US STPs	Style	Hydrazine content (ng/g)	No of replicates with peaks below LOD corresponding to hydrazine
Beech Nut	Chewing tobacco	<LOD	2
Chattanooga		<LOD	2

US STPs	Style	Hydrazine content (ng/g)	No of replicates with peaks below LOD corresponding to hydrazine
Durango		<LOD	0
Lancaster		<LOD	0
Levi Garrett		<LOD	1
Morgans		<LOD	1
Red Man Gold		<LOD	0
Red Man Regular		<LOD	0
Southern Pride		<LOD	0
Starr		<LOD	1
Stoker 707 Wintergreen		<LOD	0
Taylors Pride		<LOD	0
Trophy		<LOD	1
Bruton	Dry snuff	<LOD	2
Dental Sweet		<LOD	0
Garrett		<LOD	3

US STPs	Style	Hydrazine content (ng/g)	No of replicates with peaks below LOD corresponding to hydrazine
Honest		<LOD	3
Square		<LOD	1
CRP3		<LOD	0
Ariva Java	Hard pellet	<LOD	1
Stonewall Wintergreen		<LOD	1
Oliver Twist Original	Soft pellet	<LOD	1
Copenhagen LC	Moist snuff	<LOD	-
Copenhagen Straight LC		<LOD	0
Grizzly Natural LC		<LOD	0
Husky Natural FC		<LOD	1
Husky Straight LC		<LOD	2

US STPs	Style	Hydrazine content (ng/g)	No of replicates with peaks below LOD corresponding to hydrazine
Husky Wintergreen		<LOD	3
Kayak Straight LC		<LOD	3
Kodiak Straight LC		<LOD	1
Kodiak Wintergreen		<LOD	0
Red Seal Natural FC		<LOD	1
Red Seal Natural LC		<LOD	1
Silver Creek		<LOD	2
Skoal Straight		<LOD	1
Timber Wolf Natural FC		<LOD	0
Timber Wolf Straight LC		<LOD	0
CRP2		<LOD	2

US STPs	Style	Hydrazine content (ng/g)	No of replicates with peaks below LOD corresponding to hydrazine
Cannonball	Plug	<LOD	-
Camel Frost	US snus	<LOD	1
Camel Mellow		<LOD	0
Marlboro Mild		<LOD	1
Marlboro Peppermint		<LOD	0
Marlboro Rich		<LOD	0
Marlboro Spearmint		<LOD	1

All STPs have mean hydrazine contents below the LOD (<10 ng/g STP "as sold").

As noted above, in every case that hydrazine was tentatively identified in the current study the peak areas were substantially below the LOQ and LOD of the current method, and therefore the levels present cannot be determined. However, overall, our results indicate that hydrazine is not a prevalent contaminant of contemporary STPs, and in the minority of cases where a peak consistent with hydrazine was observed, the levels present are substantially lower than those reported previously by Liu et al. [9].

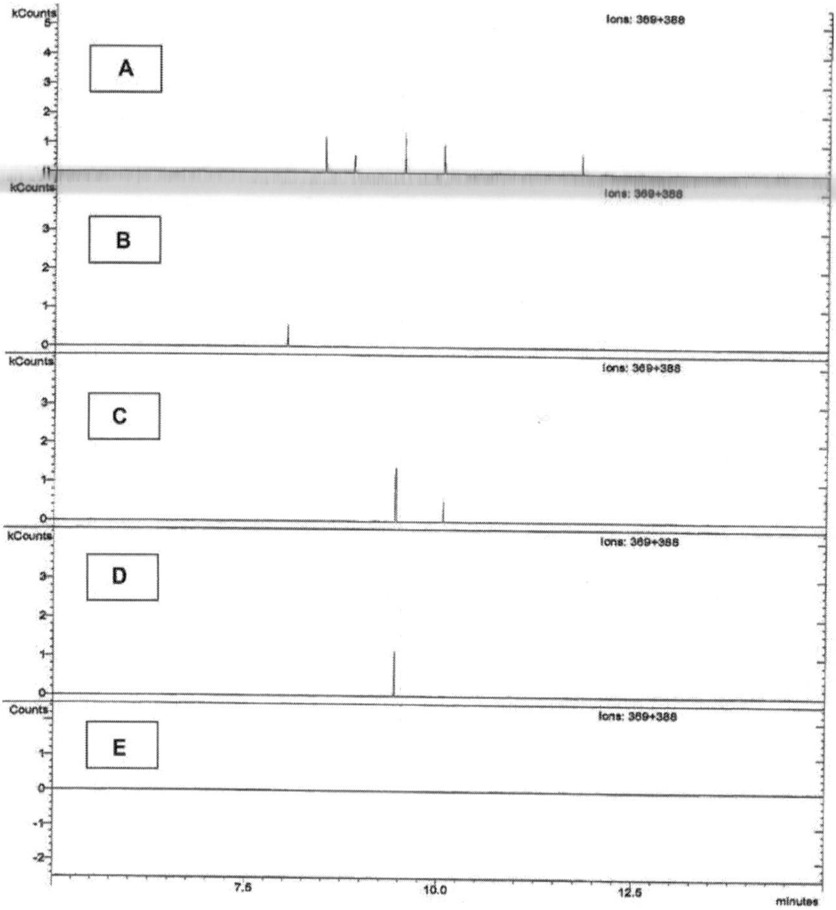

Figure 3. Typical chromatograms for smokeless tobacco products in the analysis of hydrazine (retention time 10.0 minutes); A) Camel Frost Snuff, B) Camel Mellow Snuff C) Marlboro Mild Snuff, D) Marlboro Peppermint Snuff, E) Marlboro Rich Snuff.

2.3. Sources of hydrazine in tobacco

Liu et al. [9] considered the possibility that the MH used as a sucker growth inhibitor on the tobacco crop was the source of hydrazine observed in their tobacco. Hydrazine is a contaminant in MH that derives partly from the manufacturing process and partly from subsequent breakdown of MH (particularly the formulation conjugated with diethanolamine, MH-30) [30].

However Liu et al. [9], albeit on a limited number of samples, found no relationship between MH concentrations and hydrazine. MH treated samples of tobacco had similar levels of hydrazine to samples containing no MH. Subsequent to the Liu et al. study, the diethanolamine salt was banned (in 1980) and only the more stable potassium salt of MH is currently approved for use. The US EPA [31] and the European Union [32] have also introduced limits on the concentration of hydrazine in MH - 15 ppm in the US and 1 ppm (1 μg/g) in the EU. Using the CORESTA issued Guidance Residue Levels (GRL) on agrochemicals of 80 ppm for MH on tobacco [33] as an upper limit, and assuming no hydrazine losses from the tobacco post MH-application, it can be calculated that maximum concentrations of hydrazine in tobacco arising from contaminated MH would be 1.2 ng/g in the US and 0.08 ng/g in the EU. The current study therefore does not rule out the possibility that breakdown of MH might contribute to traces of hydrazine in the tobacco.

An alternative to MH as a source of hydrazine in tobacco was advanced by Schmeltz et al. [13]. They reported the hydrazine derivative, 1,1-dimethylhydrazine, in several samples of US tobaccos using the same methodology as Liu et al. [9]. Schmeltz et al. [13] proposed that unspecified bacterial and enzymatic processes that occur during curing might be responsible for producing both the 1,1-dimethylhydrazine observed in their study and also the hydrazine observed in the earlier Liu et al. study. To date, however, there have been no reports of microorganisms or enzymatic pathways, such as nitrogenase, specifically related to tobacco that would result in hydrazone or hydrazine formation.

2.4. Differences between levels reported by Liu and results of the present study

Whatever the source of the hydrazine it is unlikely that it would be present in tobacco in a free state. Hydrazine is a powerful reducing agent, and reacts with carbonyls to form azines and hydrazones as shown in Figure 4 [34]. STPs and other forms of tobacco have been shown to contain ppm levels of formaldehyde, acetaldehyde, crotonaldehyde [10,35,36], acrolein [36], acetone, propionaldehyde, isobutyraldehyde, 2-butanone isovaleraldehyde and valeraldehyde [35]. Although the presence of carbonyl azines or hydrazones in tobacco has not been reported [37], given the thousand-fold excess of carbonyls over hydrazine in tobacco, there is a strong possibility that any hydrazine present in the tobacco plant will react with the carbonyls present. The observation that hydrazine reacts rapidly with acetaldehyde in tobacco smoke [38] adds to the plausibility that the same reaction can occur in the tobacco matrix. Moreover hydrazine is relatively volatile with a boiling point of 114°C and it would seem probable that any unreacted hydrazine would be lost during processing particularly for STPs such as snus where there is sustained heat treatment of the tobacco.

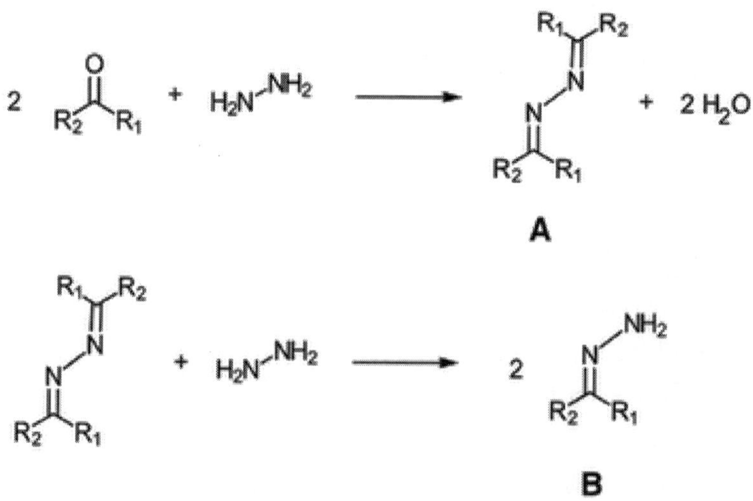

Figure 4. Reaction of ketones (R_1, R_2 = alkyl) and aldehydes (R_1 = alkyl, R_2 = H) with hydrazine to form azines (A) and hydrazones (B).

Liu et al. [9] pointed out that the highly reactive complexing agent PFB not only reacts with any free hydrazine in the matrix but is also able to react with any hydrazones or azines that may be present. They demonstrated this by showing that more than 70% of the hydrazine moiety of benzalazine was detected as pentafluorobenzaldehyde azine (PFBA) during controlled experiments over a 16 hour period representative of their experimental extraction conditions for tobacco and smoke (Figure 5). Thus we would expect that any hydrazones and azines present in the tobacco matrix could also react with PFB. Hence the hydrazine content of tobacco reported by Liu et al. may well have been the sum of hydrazine, hydrazones and azines present.

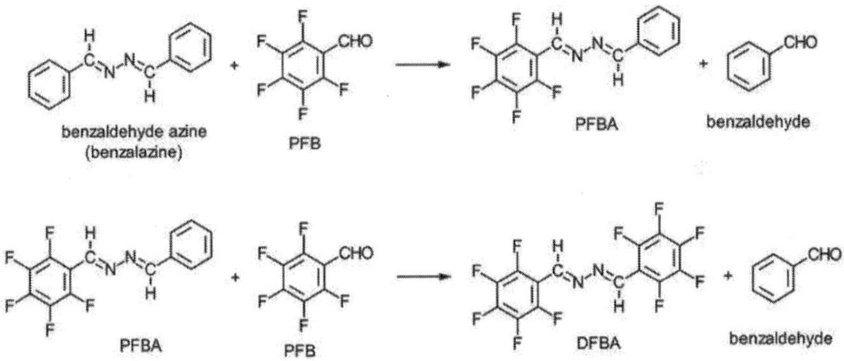

Figure 5. Reaction of benzaldehyde azine (benzalazine) with pentafluorobenzaldehyde (PFB) to form pentafluorobenzaldehyde azine (PFBA) and decafluorobenzaldehyde azine (DFBA).

In the present study a much shorter contact time between tobacco and the reactive complexing agent PFB was used. The one hour complexation time used in the present study is an order of magnitude shorter than the "overnight" time used by Liu et al. [9]. Notably, the hydrazine contents identified in this work are an order of magnitude lower than reported by Liu et al. [9]. One explanation for the difference in contents could therefore be a restricted opportunity for reaction of PFB with hydrazones or azines in this work compared to that in the study of Liu et al. [9].

2.5. Experimental

2.5.1. Tobacco samples

Tobacco samples were obtained in 2010. Details of the STP markets in the United States and Sweden were obtained, and the products for analysis were chosen to reflect approximately 90% market share of the major STP categories in these two markets at that time. The major products in each category of STP were sampled. In total, the survey comprised 31 Swedish products (10 loose snus and 21 portion snus, Table 5) and 43 US products (13 chewing tobaccos, 5 dry snuffs, 2 hard pellet products, 1 soft pellet product, 15 moist snuffs, 6 US snus and 1 plug product, Table 6). The Swedish products were sourced from Swedish retail websites, imported into the United Kingdom, and kept frozen at −20°C until analysis. The US products were sourced from shops in the United States, imported, and kept frozen at −20°C until analysis. Three CORESTA reference STP products [39] were also sampled and analysed in this exercise, CRP1 (Swedish snus pouch product), CRP2 (US style loose moist snuff) and CRP3 (US style loose dry snuff powder).

Table 5. Swedish STPs

Swedish STPs	Style	Manufacturer	Water content (%)
Ettan	Loose snus	Swedish Match	46.7
General		Swedish Match	46.4
Goteborgs Rape		Swedish Match	48.5
Granit		Fiedler & Lundgren	44.8
Grovsnus		Swedish Match	46.4
Knox		Swedish Match	43.4
Kronan		Swedish Match	48.5

Swedish STPs	Style	Manufacturer	Water content (%)
LD Original		Japan Tobacco Co.	48.0
T. Montecristo		Habanos Nordics	49.7
Skruf Strong		Skruf	49.0
Catch Licorice, mini	Portion snus	Swedish Match	42.4
Catch White Licorice		Swedish Match	46.5
CatchDry White Eucalyptus, mini		Swedish Match	22.4
Ettan		Swedish Match	42.3
General		Swedish Match	41.3
General mini		Swedish Match	44.1
General White		Swedish Match	45.3
Goteborgs Rape		Swedish Match	45.1
Granit		Fiedler & Lundgren	43.2
Granit White		Fiedler & Lundgren	39.1
Grovsnus		Swedish Match	43.9

Swedish STPs	Style	Manufacturer	Water content (%)
Grovsnus White		Swedish Match	45.3
Gustavus Original		Japan Tobacco Co.	43.0
Knox		Swedish Match	40.6
Kronan		Swedish Match	43.2
LD Original		Japan Tobacco Co.	44.0
Oomph Citrus Menthol		Northerner	4.2
Romeo y Julieta Habanos		Habanos Nordics	45.9
Skruf Strong		Skruf	36.3
Tre-Ankare White		Swedish Match	46.4
1847 Original		Philip Morris	36.8
CRP1		CORESTA Reference	44.7

Table 6 United States STPs

US STPs	Style	Manufacturer	Water content (%)
Beech Nut	Chewing tobacco	Swedish Match	21.3
Chattanooga		Swisher Int.	18.7
Durango		North Atlantic Trading Co.	20.1
Lancaster		Swisher Int.	20.2
Levi Garrett		Conwood	17.5
Morgans		Conwood	18.8
Red Man Gold		Swedish Match	21.1
Red Man Regular		Swedish Match	20.6
Southern Pride		Swedish Match	21.2
Starr		Swisher Int.	18.0
Stoker 707 Wintergreen		Swedish Match	18.7
Taylors Pride		Conwood	16.0
Trophy		Swedish Match	19.2
Bruton	Dry snuff	US Smokeless Tobacco Co.	5.8

US STPs	Style	Manufacturer	Water content (%)
Dental Sweet		Conwood	4.5
Garrett		Conwood	4.8
Honest		Conwood	4.3
Square		Swisher	7.4
CRP3		CORESTA Reference	6.8
Ariva Java	Hard pellet	Star Scientific	3.1
Stonewall Wintergreen		Star Scientific	2.7
Oliver Twist Original	Soft pellet	House of Oliver Twist	19.7
Copenhagen LC	Moist snuff	US Smokeless Tobacco Co.	47.1
Copenhagen Straight LC		US Smokeless Tobacco Co.	50.1
Grizzly Natural LC		Conwood	49.6
Husky Natural FC		US Smokeless Tobacco Co.	51.4
Husky Straight LC		US Smokeless Tobacco Co.	51.0

US STPs	Style	Manufacturer	Water content (%)
Husky Wintergreen		US Smokeless Tobacco Co.	50.3
Kayak Straight LC		Swisher	50.4
Kodiak Straight LC		Conwood	48.8
Kodiak Wintergreen		Conwood	48.0
Red Seal Natural FC		US Smokeless Tobacco Co.	49.2
Red Seal Natural LC		US Smokeless Tobacco Co.	50.1
Silver Creek		Swisher	49.5
Skoal Straight		US Smokeless Tobacco Co.	50.3
Timber Wolf Natural FC		Swedish Match	47.8
Timber Wolf Straight LC		Swedish Match	50.0
CRP2		CORESTA Reference	49.2
Cannonball	Plug	Conwood	15.4

US STPs	Style	Manufacturer	Water content (%)
Camel Frost	US snus	RJ Reynolds Co.	26.8
Camel Mellow		RJ Reynolds Co.	27.5
Marlboro Mild		Philip Morris	9.4
Marlboro Peppermint		Philip Morris	9.4
Marlboro Rich		Philip Morris	17.2
Marlboro Spearmint		Philip Morris	9.2

2.5.2. Reagents

Hexane (SpS grade) and technical grade methanol were obtained from Romil Ltd (Cambridge, UK). 1.0 N hydrochloric acid, 2,3,4,5,6-pentafluorobenzaldehyde (PFB) (purity 98%), anhydrous sodium sulphate (≥99.0%), acetic acid (purity ≥99.0%), and hydrazine sulphate (purity 99%) were obtained from Sigma-Aldrich (Gillingham, Dorset, UK). The calibration standard, decafluorobenzaldehyde azine (DFBA) (purity 99.8%), was prepared following the method of Liu et al. [9]. The purity of DFBA was characterised by GC/MS and Differential Scanning Calorimetry.

2.5.3. GC–MS conditions

GC–MS analyses were performed using a Varian 3800–Saturn 4D GC–ion trap mass spectrometer, coupled with a Varian CP-8400 autosampler and a Saturn GC–MS Workstation running Star software version 5.51 and the following conditions: column, 30 m × 0.35 mm × 0.25 μm Zebron ZB-5 capillary column; oven program, 70°C increased to 250°C at 15°C/min, followed by a 3-min dwell time (15-min run time); injection temperature, 200°C; transfer line temperature, 220°C; manifold temperature, 250°C; injection volume, 2 µl; injection, splitless; helium flow rate, 1.0 mL/min; acquisition rate, m/z 40–550; filament delay, 5 min; quantitative ions, m/z 388 + 369 (corresponding to the molecular ion $C_{14}F_{10}N_2+$ and a loss of F). An example chromatogram, for a standard solution of DFBA, is presented in Figure 6.

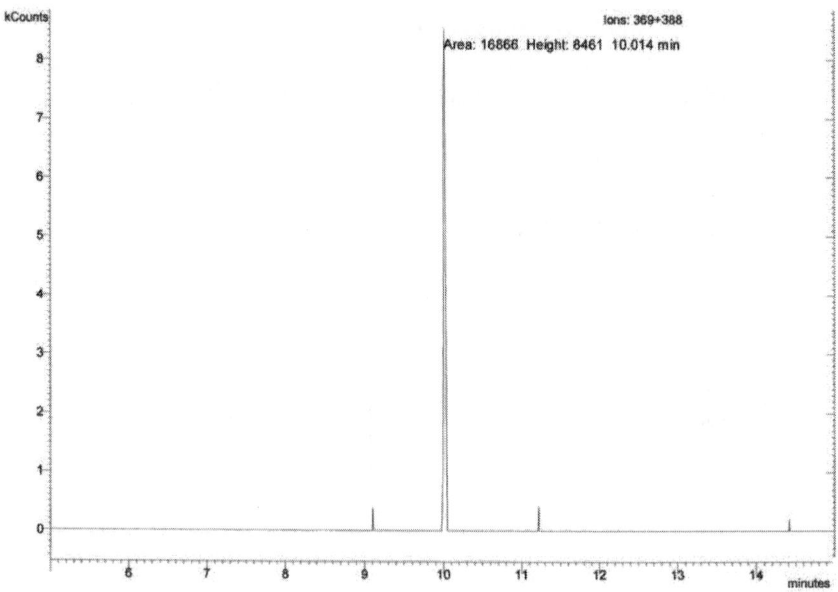

Figure 6. Typical GC–MS chromatogram for a 0.5 µg/ml decafluorobenzaldehyde azine (DFBA) standard (equivalent to 0.04 µg/ml hydrazine). Chromatogram shows signal from mass 388.

2.5.4. Preparation of standards and tobacco samples for method validation

The DFBA standard was characterized by GC–MS and the purity was determined by differential scanning calorimetry. A stock standard solution, nominally 1000 µg/mL, of DFBA in hexane, was prepared in accordance with Liu et al. [9], and diluted with hexane to give standard solutions in the range

0.20 to 2.0 μg/mL of DFBA. The solutions were tested and found to be stable at room temperature for 4 weeks.

A stock solution of hydrazine sulphate, nominally 500 μg/mL of hydrazine in deionised water was prepared for the recovery experiments. The exact concentration of hydrazine in the stock solution was calculated using the formula:

Weight of hydrazine sulphate (g)×purity of hydrazine sulphate (%)×32.05 ×106 μg.mL−1104×130.12Weight of hydrazine sulphate (g)×purity of hydrazine sulphate (%)×32.05×106 μg.mL−1104×130.12

where 32.05/130.12 is the conversion factor for hydrazine sulphate to hydrazine. The solution was stable at room temperature for 4 weeks. Diluted solutions of 1.0 to 10 μg/mL of hydrazine in deionized water were freshly prepared for recovery experiments.

Five different types of STP were used for the validation experiments: loose snus (Ettan, Swedish Match), dry snuff (Garrett, Conwood), chewing tobacco (Taylor's Pride, Conwood), hard pellet (Stonewall Wintergreen, Star Scientific), and plug (Day's Work, Swedish Match). The samples were extracted in hexane as follows. Two grams of STP were added to 50 mL of 20:80 methanol:0.1 N hydrochloric acid (aq.), and the mixture was incubated at room temperature for 1 hour. For finely divided tobacco and pellet samples, the flask was placed in an ultrasonic bath for 10 min; for leafy tobacco products the mixture was macerated with a blender for 10 min. After centrifugation for 5 min at 3000 rpm, 25 mL of supernatant was added to 0.2 mL of acetic acid and 10.0 mL of 1.0% PFB in methanol, and the mixture was swirled and allowed to stand at room temperature for 1 hour. Next, 20 mL of hexane was added, and the mixture shaken for 1 min. The aqueous phase was run into a clean flask, and the hexane fraction was passed over anhydrous sodium sulphate. The aqueous phase was extracted again by the above procedure with 20 mL of hexane, and then again with 10 mL of hexane, and the three extracts were combined. The combined hexane extract was reduced to approximately 1 mL using a rotary evaporator, transferred to a 2-mL volumetric flask, and made up to 2 mL with fresh hexane, ready for application to GC–MS.

The amount of hydrazine in each tobacco sample was calculated using the formula:

Test sample area×std. conc. (μg.mL−1)×50×2×32.05×1000Mean std. area×2 ×25×388.17Test sample area×std. conc. (μg.mL−1)×50×2×32.05×1000Mean std. area×2×25×388.17

where 32.05/388.17 is the conversion factor for DFBA to hydrazine. Values in this study are reported in units of ng/g, as the common portion sizes of use are gram sized [40].

For recovery experiments, 2 g of STP was added to 100 μl of a standard hydrazine solution (0.5 μg/g, 0.05 μg/g and 0.025 μg/g hydrazine). The mixture was swirled and allowed to stand at room temperature for 10 min to enable the hydrazine to penetrate the matrix. Next, 50 mL of 20:80 methanol:0.1 N hydrochloric acid (aq.) was added and the tobacco sample extracted as described above. Accuracy was determined as the mean recovery ± relative standard deviation (RSD). Precision was determined as the repeatability RSD at each level.

2.5.5. Water content

STP samples were analysed for water content using Karl Fisher Coulometric analysis with an KEM MKC-500 analyser (Kyoto Electronics, Tokyo, Japan). Approximately 2 g STP was weighed into a 25 ml snap-top vial. 20.0 ml MeOH was added and the sample sonicated for 15 minutes before being allowed to steep and settle for at least 2 hours. 100 μl of methanol was sampled and injected into the Karl Fisher analysis cell. Water blanks were subtracted, and analyses conducted in triplicate.

3. CONCLUSIONS

In this study we have developed and validated a method to determine levels of hydrazine in tobacco, using PFB as a complexing agent to form the azine, DFBA, which was then quantified by GC-MS. In a series of 74 brands of smokeless tobacco from the US and Sweden covering all major product styles - snus, chewing tobacco, moist snuff, dry snuff, plug and pellet products - hydrazine concentrations were all below the level of quantification of 26.5 ng/g product. Peaks consistent with hydrazine were identified, at trace levels (below limit of detection), in the chromatograms of less than half of the 74 STPs, but at levels considerably lower than the only previous study to have quantified hydrazine in tobacco. All previous references to the presence of hydrazine in tobacco and STPs are based on a single study from 1974 [9] which measured the hydrazine contents of tobacco from five different cigarettes (4 experimental and 1 commercial).

There is insufficient information to explain the differences in results between the present study and the earlier study. However the much lower levels of hydrazine found in the present study are consistent with the reductions in maximum levels of hydrazine residues permitted in MH, which were legislated subsequent to the original study of Liu et al. Also our method used shorter times for complexation of hydrazine residues with PFB. The shorter complexation time may have minimised artifactual formation of DFBA through complexation with non-hydrazine moieties such as hydrazones and azines as Liu et al. reported as probable in their study. Nevertheless, the current study demonstrates that the presence of hydrazine in contemporary STPs is relatively infrequent, and when present hydrazine is at levels below the limits of quantification and detection that can be achieved using current analytical best-practice.

ACKNOWLEDGEMENTS

We would like to acknowledge the contributions of GC Laboratories Ltd., Hitchin, UK to method development and sample analysis activities.

COMPETING INTERESTS

The study was funded by British American Tobacco (BAT). KM, HK, CV, PD and CW are employees of BAT, as was SE at the time of the study. AP is a paid consultant to BAT. BR's research is part funded by unrestricted grants from Tobacco Manufacturers (including BAT) to the University of Louisville.

AUTHORS' CONTRIBUTIONS

KM – co-directed the study and co-wrote the manuscript. HK – project managed the study. AP - co-wrote the manuscript. CV – identified, sourced and summarised hydrazine literature. SE, PD, CW – guided and managed the method development and validation. BR - co-directed the study and contributed to writing the manuscript.

REFERENCES

1. IARC. Smokeless tobacco and some tobacco-specific N-nitrosamines. IARC Monographs on the Evaluation of Carcinogenic Risks to Humans, IARC Press, Lyon, France. 2007;89:370.

2. Royal College of Physicians. Harm reduction in nicotine addiction: helping people who can't quit. London: RCP: A report by the Tobacco Advisory Group of the Royal College of Physicians; 2007.

3. FDA: Draft guidance for industry: reporting harmful and potentially harmful constituents in tobacco products and tobacco smoke under the Federal Food, Drug, and Cosmetic Act. 2012. [http://www.fda.gov/TobaccoProducts/GuidanceComplianceRegulatoryInformation/ucm297786.htm]

4. IARC. Re-evaluation of some organic chemicals, hydrazine and hydrogen peroxide. IARC Monographs on the Evaluation of Carcinogenic Risks to Humans, IARC Press, Lyon, France. 1999;71:991–1013.

5. Environment Canada/Health Canada. Screening Assessment for the Challenge. Hydrazine. Chemical Abstracts Service Registry Number 302-01-2. January 2011. http://www.ec.gc.ca/ese-ees/17647095-B851-46F4-A4BB-79F887D84666/batch10_302-01-2_en.pdf

6. Toth B. Synthetic and naturally occurring hydrazines as possible cancer causative agents. Cancer Res. 1975;35:3693–7.

7. Thorneley RNF, Eady RR, Lowe DJ. Biological nitrogen fixation by way of an enzyme-bound dinitrogen-hydride intermediate. Nature. 1978;272:557–8.

8. Dilworth MJ, Eady RR. Hydrazine is a product of dinitrogen reduction by the vanadium-nitrogenase from Azotobacter chroococcum. Biochem J. 1991;277:465–8.

9. Liu YY, Schmeltz I, Hoffmann D. Chemical studies on tobacco smoke. XXVII. Quantitative analysis of hydrazine in tobacco and cigarette smoke. Anal Chem. 1974;46:885–9.

10. Hoffmann D, Adams JD, Lisk D, Fisenne I, Brunnemann KD. Toxic and carcinogenic agents in dry and moist snuff. J Natl Cancer Inst. 1987;79:1281–6.

11. Brunnemann KD, Hoffmann D. Chemical composition of smokeless tobacco products. In Smoking and Tobacco Control Monograph. 1993;2:96–108.

12. Schmeltz I, Hoffmann D. Nitrogen-containing compounds in tobacco and tobacco smoke. Chem Rev. 1977;77:295–311.

13. Schmeltz I, Abidi S, Hoffman D. Tumorigenic agents in unburned processed tobacco: N-nitrosodiethanolamine and 1,1-dimethylhydrazine. Cancer Lett. 1977;2:125–32.

14. Hecht SS. Cigarette smoking and cancer. In: Rom WN, editor. Environmental and occupational medicine. New York: Lippincott-Raven; 1998. p. 1479–99.

15. Hatsukami DK, Ebbert JO, Feuer RM, Stepanov I, Hecht SS. Changing smokeless tobacco products. New tobacco-delivery systems. Am J Prev Med. 2007;33 Suppl 6:S368–78.

16. Scientific Committee on Emerging and Newly Identified Health Risks. Health effects of smokeless tobacco products. 2008 http://ec.europa.eu/health/archive/ph_risk/committees/04_scenihr/docs/scenihr_o_013.pdf

17. Patterson JM, Issidorides CH, Groutas VC, Smith Jr WT. Pyrolysis of maleic hydrazide and MN'dimethyldodecylamine. London: Chem Ind (London); 1972. p. 337–8.

18. Pailer M, Sulm O: Pyrolyse von Tabakadditiven. Fachliche Mitt Oesterr Tabakregie. 1973;14:258–266.

19. Harke H-P, Schueller D, Drews C-J. Pyrolyse von maleinsäurehydrazid und hydrazin. Z Lebensm Unters-Forsch. 1973;153:163–9.

20. Elias G, Bauer WF. Hydrazine determination in sludge samples by high-performance liquid chromatography. J Sep Sci. 2006;29:460–4.

21. Kirchherr H. Determination of hydrazine in human plasma by high-performance liquid chromatography. J Chromatogr. 1993;617:157–62.

22. Khuhawar MY, Rajper AD, Rind FMA. High performance liquid chromatographic determination of hydrazine in water using 2-hydroxynaphthaldehyde as a derivatizing reagent. Pak J Sci Ind Res. 2002;45:371–4.

23. George M, Nagaraja KS, Balasubramanian N. Spectrophotometric determination of hydrazine. Talanta. 2008;75:27–31.

24. Ortega-Barrales P, Molina-Díaz A, Pascual-Reguera MI, Capitán-Vallvey LF. Solid-phase spectrophotometric determination of trace amounts of hydrazine at sub-ng ml^{-1} level. Anal Chim Acta. 1997;353:115–22.

25. Sun M, Bai L, Liu DQ. A generic approach for the determination of trace hydrazine in drug substances using in situ derivatization-headspace GC–MS. J Pharm Biomed Anal. 2009;49:529–33.

26. Haji Shabani AM, Dadfarnia S, Dehghan K. Indirect spectrophotometric determination of trace quantities of hydrazine. Bull Korean Chem Soc. 2004;2004(25):213–5.

27. George M, Nagaraja KS, Balasubramanian N. Spectrophotometric determination of hydrazine using bromine and methyl red. Indian J Chem. 2007;46A:1621–4.

28. McAdam K, Faizi A, Kimpton H, Porter A, Rodu B. Polycyclic aromatic hydrocarbons in US and Swedish smokeless tobacco products. Chemistry Central. 2013;7:151.

29. The Codex Alimentarius Volume 3. Residues of veterinary drugs in foods. 2nd ed. Rome Italy: Joint FAP/WHO Food Standards Programme, FAO; 1993. p. 59.

30. Bakker H, Martyn A, Schreuder RH. Gas–liquid chromatographic determination of hydrazine in maleic hydrazide formulations and in samples stored at an elevated temperature. Pestic Sci. 1983;14:470–4.

31. US Environmental Protection Agency. Maleic hydrazide: Notification of issuances of notice of intent to suspend pesticide registration. Document OPP38501; PH-FRL-1934. Fed Regist. 1981;46(179):45999–6000.

32. European Commission: Health & Consumer Protection Directorate-General: Review report for the active substance maleic hydrazide. SANCO/10501/2002-final 2 December 2002. http://ec.europa.eu/food/plant/protection/evaluation/existactive/list1-40_en.pdf

33. CORESTA, 2010, http://www.coresta.org/Guides/Guide-No01-GRLs(3rd-Issue-July13).pdf

34. March J: Advanced Organic Chemistry 4th Edition, Hoboken, New Jersey: Wiley Interscience; 1992, p.904.

35. Weybrew JA, Stephens RL. Survey of the carbonyl contents of tobacco. Tob Sci. 1962;6:53–7.

36. Stepanov I, Jensen J, Hatsukami D, Hecht SS. New and traditional smokeless tobacco: comparison of toxicant and carcinogen levels. Nicotine Tob Res. 2008;10:1773–82.

37. Rodgman A, Perfetti TA. The chemical components of tobacco and tobacco Smoke. Second Edition: CRC Press; 2013.

38. Harward CN, Parrish ME, Plunkett SE, Banyasz JL, Shafer KH. Evaluation of hydrazine reduction by cellulose acetate filters using infrared tunable diode laser spectroscopy. Anal Chem. 2002;74:5871–81.

39. CORESTA, 2010, http://www.coresta.org/Reports/STS-CTR_2010-4CRP-Analysis.pdf

40. Digard H, Errington G, Richter A, McAdam K. Patterns and behaviors of snus consumption in Sweden. Nicotine Tob Res. 2009;11:1175–81.

CHAPTER 7

Determination of 8 Diuretics and Probenecid in Human Urine by Gas Chromatography-Mass Spectrometry: Confirmation Procedure

Olga Zaporozhets, Iuna Tsyrulneva, Mykola Ischenko

Taras Shevchenko National University of Kiev, Kiev, Ukraine

ABSTRACT

A fast and sensitive method for determination of 8 diuretics (acetazolamide, bendroflu-methiazide, bumetanide, chlorthalidone, furosemide, hydrochlorothiazide, metolazone, triamterene) and masking agent (probenecid) in human urine using gas-chromatography with mass spectrometric detection is described. The extraction of the substances as function of the nature of organic solvent, mixing time and pH of aqueous phase was studied. The tandem mass spectrometry was used to increase selectivity of diuretics determination due to elimination of background interferences. Fragmentation reactions were studied for each compound and their collision energies were optimized to obtain the best selectivity. The results of method's validation demonstrate its suitability in routine analysis for confirmation purposes.

KEYWORDS

Diuretics; Urine; Gas Chromatography; Tandem Mass Spectrometry; Confirmation

1. INTRODUCTION

It is known [1] that athletes misuse diuretics for several reasons: to reduce body weight in order to be qualified for a lower weight category, to decrease fluid retention in body caused by the application of steroids and to reduce the concentration of other prohibited substances in urine to avoid positive doping results. International Olympic Committee and World Anti-Doping Agency (WADA) prohibited the use of diuretics in sport. WADA established the requirement for anti-doping laboratories to control the presence of these compounds in urine at the level less or equal to minimum required performance limit (MRPL), which is 0.25 $\mu g.ml^{-1}$ [2]. Most of the banned substances should not be present at any concentration in the specimens (mainly urine) collected for the analysis. Therefore, they are often detected only on a qualitative basis.

Diuretics belong to non-threshold substances, which means the absence of specific threshold concentration in urine. **Table 1** shows that diuretics include compounds which have differences in molecular structures and chemical properties [3], and this fact complicates the elaboration of the only method of their extracting from biological matrix and their detection at the necessary selectivity level.

According to WADA requirements the only methods of analysis that should be applied at anti-doping laboratories are liquid or gas chromatography with mass spectrometric detection. Though liquid chromatography with mass spectrometric detection (LC-MS) causes the decrease in sensitivity due to the suppression of ionization by biological matrix [4,5] and shift in retention times in case of direct injection of sample [6] it also allows to achieve low detection limits and short time of analysis [7]. Because of relatively high cost of LC-MS this method is more preferable for quantative determination of diuretics in biological objects and for screening purpose. Gas chromatography with mass detection (GC-MS) is suitable for multicomponent, selective and sensitive diuretics determination in human urine, as well as liquid chromatography. GC-MS method which includes two-step liquid-liquid extraction and derivatization was proposed for quantative determination of 16 diuretics [8]. Long preliminary sample preparation of urine limits the use of the method in routine analysis. Moreover, this technique does not provide high recovery of chlorthalidone and triamterene. At the same time GC-MS demonstrates the best results for qualitative determination and for confirmation.

The GC-MS method was proposed for confirmation in [9]. The recovery of diuretics and impeding effect of urine matrix and other diuretics were not studied in this work as well as in [10]. Besides, both methods were not validated. Moreover, the technique proposed in [11] is long lasting which is caused by long retention time.

Despite the great number of methods proposed, gas or liquid chromatography with mass spectrometric detection do not always provide the necessary selectivity towards the urine components [10,12]. In case of liquid chromatography, the use of tandem mass-spectrometry allowed to pass over these problems and to ensure the satisfactory result [13-15].

Despite liquid chromatography is more preferable for screening purposes, gas chromatography with mass detection demonstrates the best results for confirmation. According to WADA requirements the optimal method for confirmation of presence of diuretics is the method different from the screening method. Therefore, the goal of the present work was to develop a method of confirmation of presence of diuretics and probenecid in urine by gas chromatography with tandem mass spectrometric detection.

Table 1. General characteristics of diuretics.

Hydrochlorothiazide	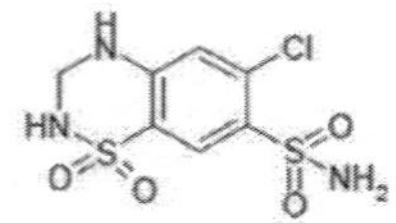	7.9	9.2	Bumetanide	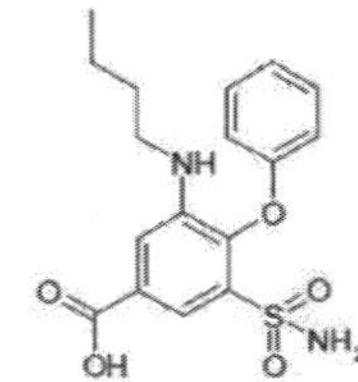	3.6	7.7
Trichlormethiazide	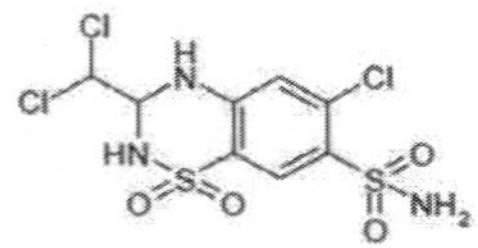	7	10	Probenecid	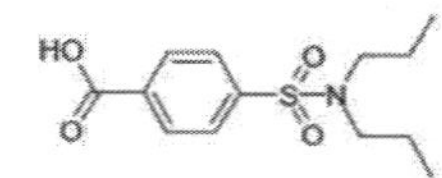	3.4	-
Bendroflumethiazide	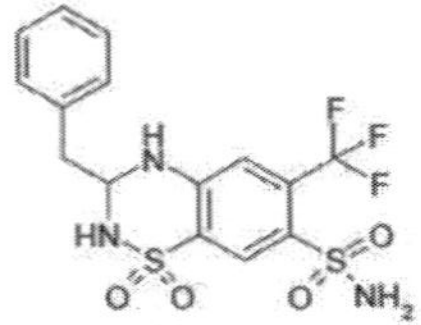	8.5	-	Ethacrinic acid (Internal Standard)	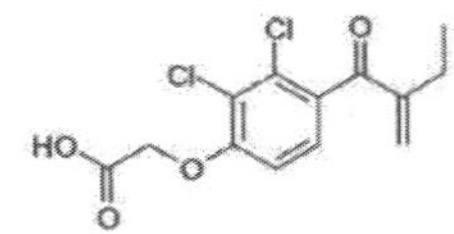	3.5	-
Metolazone	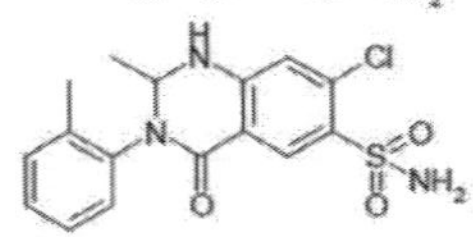	9.7	-				

Name	Structure	pK_a		Name	Structure	pK_a	
		pK_{a1}	pK_{a2}			pK_{a1}	pK_{a2}
Triamterene	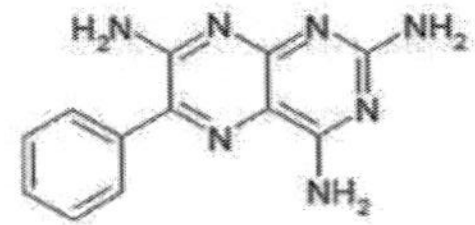	6.2	-	Indapamide	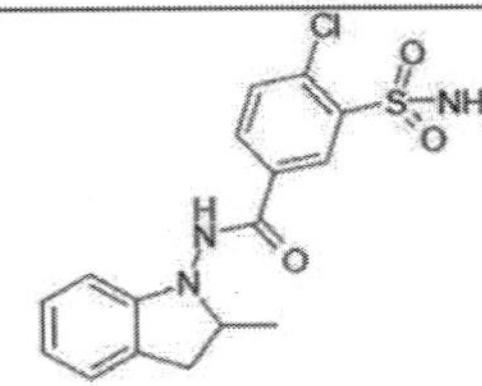	8.8	-
Acetazolamide	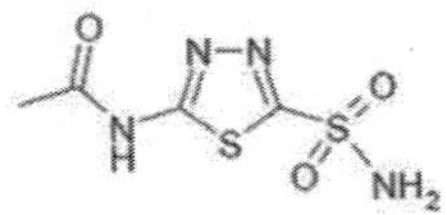	7.4	9.1	Chlorthalidone	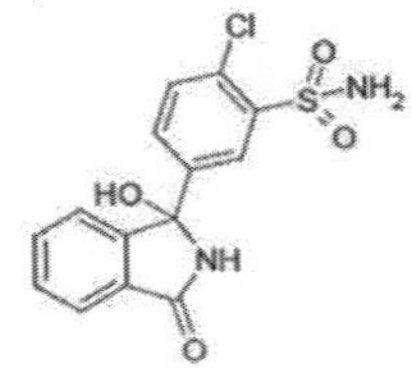	9.4	-
Chlorothiazide	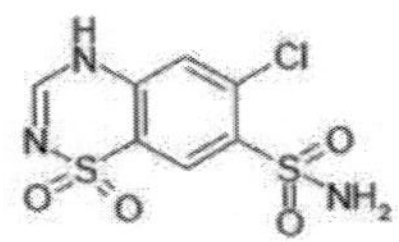	6.7	9.5	Furosemide	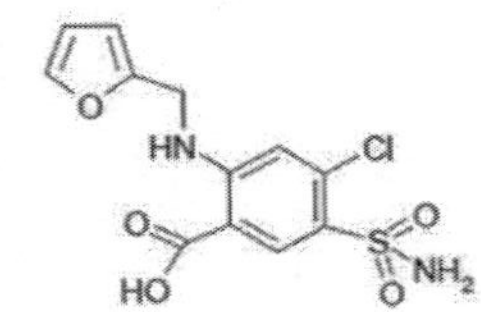	3.9	-

2. EXPERIMENTAL PROCEDURE

2.1. Preparation of Reagent Solutions

Stock standard solutions were prepared for 12 diuretic drugs (acetazolamide, bendroflumethiazide, bumetanide, chlorthalidone, chlorothiazide, ethacrinic acid, furosemide, hydrochlorothiazide, indapamide, metolazone, triamterene, trichlormethiazide) and a masking agent (probenecid) (all by SIGMA, St. Louis, MO, USA) by dissolving in methanol at concentration of 1.0 mg.ml^{-1}.

Diethyl ether, ethyl acetate, isopropanol, acetone (Merck, Darmstadt, Germany) and methanol (SIGMA, St. Louis, MO, USA) had a qualification "for chromatography". Methyl iodide with purity 99% (SIGMA, St. Louis, MO, USA), sodium sulfate (Merck, Darmstadt, Germany) were used without further purification, potassium carbonate (Merck, Darmstadt, Germany) was used after calcination in oven. Hydrochloric acid and sodium hydroxide were prepared by dilution of the concentrated initial solution (Merck, Darmstadt, Germany).

2.2. Instruments and Materials

Gas chromatograph Varian 3800 (Australia) equipped with an automatic sample injector system Combipal and FactorFour Capillary Column VF-DA 12 m, 0.20 mm I.D. for mass spectrometric determination of narcotic drugs and pharmaceuticals. Mass spectrometer Varian 1200 (Australia), directed by computer system of data processing Varian MsWorkstation 6.9, with a system of three quadruples. Auxiliary equipment: thermostat VLM EC1 (Germany), evaporator in nitrogen stream Liebisch (Germany), centrifuge OPN-3Y4.2 (Russia), shaker Heidolph Vibramax 110 (Germany), oven Binder (Germany), pHmeter inoLab pH level 1 (Germany), analytical balance KERN ABS (Germany), pipettes of adjustable volume Eppendorf (Germany) 0 - 20 μl, 20 - 200 μl, 100 - 1000 μl, 500 - 5000 μl.

2.3. Instrumental Parameters

Instrumental parameters of chromatographic determination: carrier gas—helium, flow rate—1 ml.min^{-1}, injector temperature—290°C, injector mode—split 20; volume of sample injected—2 μl; flushing liquid—acetone, source temperature—225°C, interface temperature—250°C; SIM Width—0.7; Scan Time—0.2 s; start of the ion current registration—5.3 min. Instrumental parameters of tandem mass spectrometry: auxiliary gas pressure (argon) on the second quadruple—2.10 - 1.65 mTorr, electron energy at the source—20 eV. Registration mode: selected ion monitoring (SIM) or multiple reaction monitoring (MRM). In experiment mass spectrometry detection was carried out in SIM mode using ions listed in **Table 2** [12].

2.4. Sample Preparation

Negative urine samples were taken from volunteers who did not consume banned substances; the positive (spiked) urine samples were prepared by adding mixtures of standard diuretics solutions to the negative urine samples prior to the sample preparation. The urine samples were stored in polypropylene bottles at a temperature of –20°C. Before conducting the experiment samples were defreezed, centrifuged for 10 min at 1500 rpm, and the pH value was measured. Ethacrinic acid (internal standard, IS), mixture of standard diuretics, 200 mg of Na_2SO_4 and 5.0 ml of urine were transferred in a 12 ml glass tube. The pH value was adjusted by adding NaOH or buffer solution and 5.0 ml of organic solvent. The tubes were shaken in a shaker for 5 - 20 min and then centrifuged. The top layer was totally transferred to a 10 ml glass tubes. Solutions were evaporated in nitrogen stream at 40°C (for solvents based on diethyl ether) or 80°C (for solvents based on ethyl acetate) to obtain solid residue, the tubes were cooled to room temperature. The derivatization (methylation by methyl iodide in acetone in an alkaline environment) was performed according to [16]. Afterwards the samples were immediately sent to gaschromatographic analysis. The degree of substances recovery from matrix achieved by liquid-liquid extraction was investigated at a concentration of 0.9 MRPL (0.225 μg.ml^{-1}). The mixture of standard diuretics and IS were added to the urine samples and sample preparation was performed by the method described. Simultaneously, the derivatization of mixture of diuretics standards with the same concentration was conducted. All samples were analyzed by chromatographic-mass spectrometric method. The recovery degree was calculated by taking the ratio of peak height obtained at the retention time for the analyte of interest compared to that obtained from an IS in spiked urine and mixture of standards.

2.5. Method Validation

To determine all parameters (capability to identification, selectivity, detection limit), that are necessary for suitability of verification in accordance with ISO/IEC 17025: 2006 [17], the investigation was performed. Linearity, accuracy and precision were not examined since the method is intended for confirmation and is not intended for quantitative determination of diuretics, as established in [17,18]. Therefore more consideration was given to definite identification of the analyte in urine.

Identification by the retention time. A series of spiked urine samples and mixture of standards were simultaneously analyzed to check the capability of identification. Absolute and relative differences in retention times were calculated.

Mass spectrometry identification in tandem mass spectrometry (MS-MS) mode. A series of extracts of spiked urine samples and mixture of standards at concentrations 0.5, 1.0 and 5.0 MRPL were simultaneously analyzed to prove the possibility of mass spectrometric identification of the analytes. The relative intensity of the diagnostic ion was calculated on the base of the chromatographic peaks' area or height, integrated by corresponding ion.

The selectivity of method was tested by analyzing six different urine samples after sample preparation without analytes.

The limit of detection was evaluated according to [19]. The samples were prepared by addition of the mixture of standards to urine to achieve the concentrations 0.2, 0.5, 1.0, 2.0 MRPL, and analyzed by marking the number of positive determinations regarding to total number of determinations.

That concentration level, which corresponded to an unambiguous detection in all the samples, was considered as a limit of detection.

Table 2. Characteristic ions of diuretics in Selected Ion Monitoring mode.

Analyte	Retention time, min	Segment range, min	Ions, m/z
Triamterene	7.46	7.2 - 7.9	336
Acetazolamide	4.46	4.2 - 4.6	249, 108
Chlorothiazide	6.46	5.5 - 6.82	248, 275
Hydrochlorothiazide	7.38	7.2 - 7.9	310
Trichlormethiazide	8.11	7.9 - 8.5	352, 354
Bendroflumethiazide	8.73	8.5 - 10.0	386
Metolazone	10.87	10.0 - 12.0	392
Indapamide	7.46	7.2 - 7.9	407
Chlorthalidone	6.94	6.82 - 7.2	287, 363
Furosemide	6.68	5.5 - 6.82	81, 372
Bumetanide	7.05	6.82 - 7.2	406, 363
Probenecid	4.77	4.6 - 5.5	270
Ethacrinic acid	4.91	4.6 - 5.5	261

2.6. Quality Control

While carrying out the experiment, the mixture of diuretics standards at 0.9 MRPL (standard sample), the urine sample obtained from a volunteer who did not consume prohibited substances (negative sample), spiked urine sample with concentration equal to 0.9 MRPL (positive sample) and reagent blank (distilled water) were also analyzed using the same method of sample preparation for each analytical run. All samples, except for the reagent blank, were prepared with the addition of IS.

3. RESULTS AND DISCUSSION

3.1. Optimization of Gas-Chromatographic Method

The method of diuretics determination includes their extraction by organic solvents, subsequent derivatization and following chromatographic separation and detection. In this work the recovery degree as the function of the nature of organic solvent, mixing time and pH of aqueous phase were investigated to optimize the conditions of diuretics determination.

The liquid-liquid extraction of diuretics from urine was carried out with the use of diethyl ether, ethyl acetate and mixtures of organic solvents (ratio): diethyl ether and isopropanol (19:1), diethyl ether and methanol (19:1), ethyl acetate and isopropanol with different ratio of components (19:1, 9:1, 17:3). The obtained results show (**Figure 1**) that the higher recovery degree for most of diuretics has been achieved at their extraction from mixture ethyl acetate:isopropanol = 17:3. The extraction degree of all substances is satisfactory, though the recovery of metolazone and trichlormethiazide under these conditions is 53.7% and 3.1%, respectively, but the signal intensity is sufficient enough to detect these compounds at the MRPL level. The extraction equilibrium in the systems is reached for 5 minutes at given conditions.

The diuretics can be classified as weak acids (carboanhydrase inhibitors, such as acetazolamide; thiazide derivatives and those that belong to them, for example, chlorthalidone), strong acids (loop diuretics such as furosemide) and base compounds (triamterene). Considering this fact extraction of selected diuretics as the function of pH value of the aqueous phase was investigated. It was found experimentally that weakly acidic diuretics are effectively extracted in molecular form in the range of pH 6 - 8 (pH of urine is ~5.5 - 7.0). The degree of their extraction exceeds 50%. Extraction degree of triamterene also increases. Extraction degree of more acidic diuretics such as furosemide, bumetanide and probenecid is lower under these conditions, but high chromatographic yield of characteristic ions of these compounds allows to detect them at the MRPL level. Further the acidity of aqueous phase was adjusted by NaOH addition to achieve pH 6 - 8.

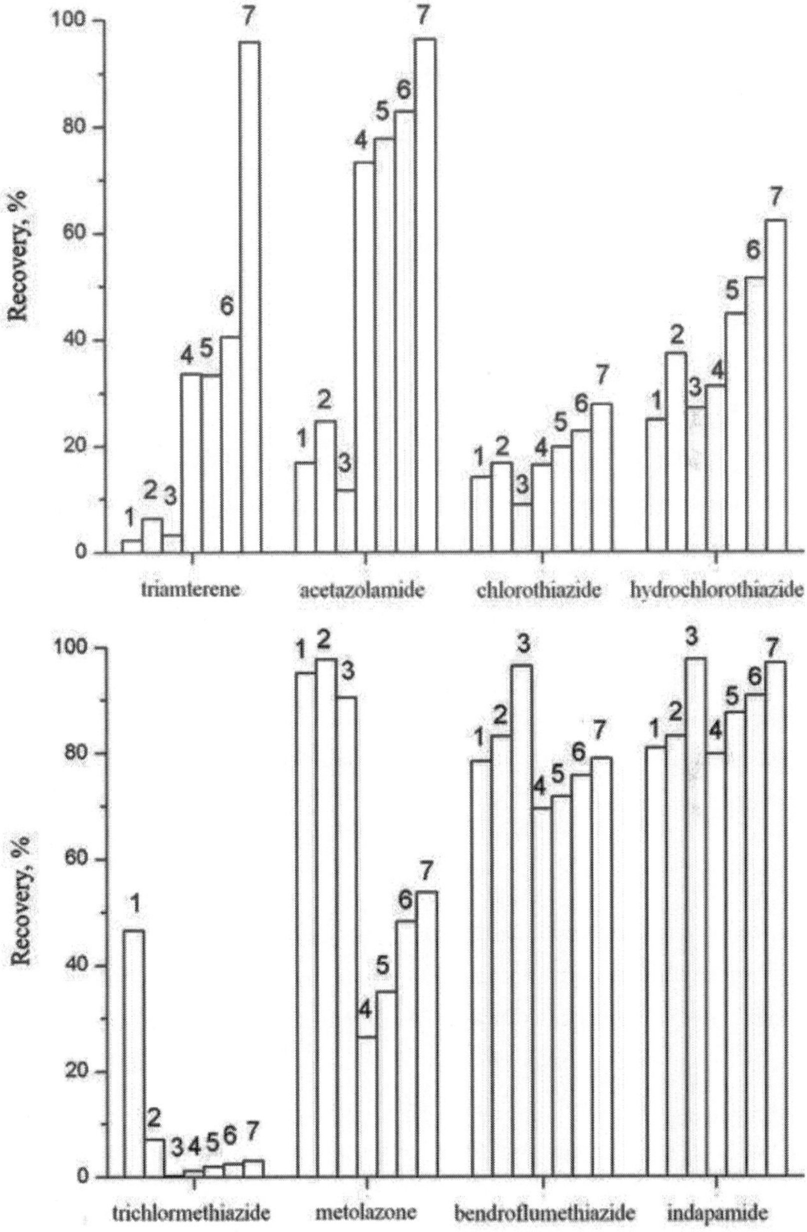
Recovery, %
triamterene
acetazolamide
chlorothiazide
hydrochlorothiazide
trichlormethiazide
metolazone
bendroflumethiazide
indapamide

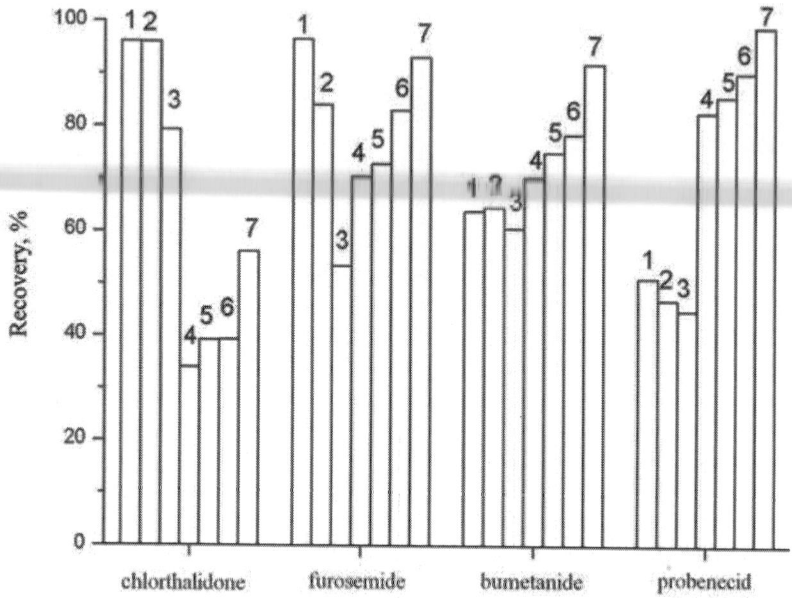

Figure 1. The recovery of diuretics as function of solvents. V_{solv} = 5 ml, V_{aq} = 5 ml, pH = 7.0, T = 295 ± 1 K. 1: Diethyl ether; 2: Diethyl ether:Isopropanol = 19:1; 3: Diethyl ether: Methanol = 19:1; 4: Ethylacetate; 5: Ethylacetate:Isopropanol = 19:1; 6: Ethylacetate:Isopropanol = 9:1; 7: Ethylacetate:Isopropanol = 17:3.

Considering the demands made for chromatographic column, such as providing maximum separation of analytes in a short time and compatibility with mass detector, capillary polymethacrilate column FactorFour VF-DA 12 m, 0.20 mm I.D., which was successfully used for determination of narcotic drugs and pharmaceuticals, was chosen. To increase the efficiency of chromatographic separation of diuretics and to provide high rate of the analysis the optimization of temperature program was carried out. The following parameters were varied: initial and final temperature, heating rate and held. The optimal temperature program that meets all the requirements has following parameters: initial and final temperatures 120 and 300°C, respectively; first held for 0.5 min, followed by heating rate 30°C.min^{-1} and subsequent held for 8.5 min. The chromatogram of diuretics standards mixture under optimal conditions is shown in **Figure 2**. It is seen that complete chromatographic separation of hydrochlorothiazide, triamterene and indapamide at this temperature program is not achieved. However, these compounds can be easily discriminated due to different mass spectral features of diuretics. Besides, the presence of various diuretics in the same sample is unlikely. Time of chromatography is 12 min under the optimal conditions.

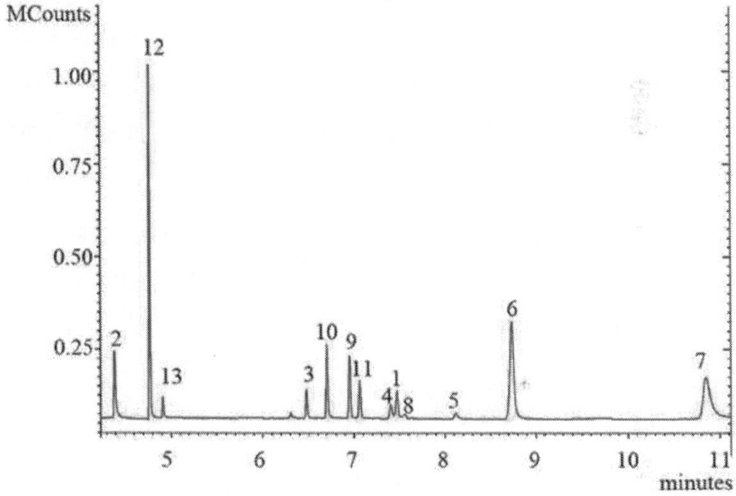

Figure 2. The chromatogram of the diuretics standard mixture, analyzed in SIM acquisition mode. C = 0.225 μg.ml^{-1}. 1: Triamterene; 2: Acetazolamide; 3: Chlorothiazide; 4: Hydrochlorothiazide; 5: Trichlormethiazide; 6: Bendroflumethiazide; 7: Metolazone; 8: Indapamide; 9: Chlorthalidone, 10: Furosemide; 11: Bumetanide; 12: Probenecid; 13: Ethacrinic acid.

3.2. Optimization of MS-MS Conditions

One of the major disadvantages of mass spectrometric methods is low selectivity in the analysis of biological objects. The identification of analytes in biological matrices is complicated due to misrepresentation of mass spectrum caused by concomitant substances. It can lead to appearance of false-positive results. In this case tandem mass spectrometry is more preferable method [20]. The combination of mass selection of the precursor followed by potentially unique collision-induced dissociation and mass selection or scanning of the product ion provides higher specificity, since coincidence of spectrum of dissociated ion and ion with known structure is a conclusive proof of ions structure identity. The tandem mass spectrometry method can be used either in full scan acquisition mode or multiply reaction monitoring mode. The second technique is more sensitive in comparison with the first one. Typically, two transitions precursorproduct must be monitored, but in some cases the combination of a single precursor-product ion pair is sufficiently unique to be definitive.

For each diuretic molecular ions, product ions and collision energy (E) were chosen. Collision energy should ensure maximum yield of product ions [21]. The data for transition reaction molecular ion—product ion are presented in **Table 3**.

Chromatograms of spiked with bendroflumethiazide and triamterene chosen as examples urine samples and negative urine samples at these conditions are presented in **Figure 3**.

It is shown that the compounds may be detected in each of the reactions. Besides, peaks with signal-to-noise ratio ≥ 3 at the time which corresponds to analyte retention time, are absent in chromatogram of negative sample. This result was observed for other diuretics and probenecid as well. This fact proves sufficient selectivity of method.

3.3. Method Validation

According to the requirements of WADA [21] the relative intensities of any of the ions in mass shall not differ by more than the amount in **Table 4** from the relative intensities of the same ions acquired from spiked urine.

The retention times of all diuretics (except metolazone) do not differ by more than 1% from that of the same substance in a spiked urine sample. However, there are no peaks with signal-to-noise ratio ≥ 3 in chromatograms of six analyzed blanks for the following compounds: acetazolamide, furosemide, bumetanide, indapamide, chlorothiazide, hydrochlorothiazide, metolazone, bendroflumethiazide, trichlormethiazide, chlorthalidone, triamterene.

Table 3. The diagnostic reactions and optimal collision energies for diuretics determination in Multiple Reaction Monitoring mode.

Analyte (molecular ion)	Reaction, m/z	E, eV	Reaction, m/z	E, eV	Reaction, m/z	E, eV
Triamterene (336)	336→336[a]	–10	336→320	–23	336→292	–20
Acetazolamide (249)	249→249[a]	–6.5	249→83	–12	249→108	–13
Chlorothiazide (248)	248→248	–6.5	248→141[a]	–10	248→140	–28
Hydrochlorothiazide (353)	353→353	–6.5	353→288[a]	–10	353→180	–22
Trichlormethiazide (399)	399→399	–6.5	399→192[a]	–28	399→90	–15
Bendroflumethiazide (386)	386→386[a]	–8	386→278	–24	386→279	–17
Metolazone (392)	392→392[a]	–15	392→285	–23	392→284	–25
Indapamide (407)	407→407	–5	407→161[a]	–21	407→132	–10
Chlorthalidone (287)	287→287[a]	–8	287→255	–15	287→220	–20
Furosemide (372)	372→372	–5	372→96	–12	372→81[a]	–17
Bumetanide (406)	406→406[a]	–6	406→363	–10	406→318	–17
Probenecid (270)	270→270[a]	–5	270→199	–10	270→135	–15
Ethacrinic acid, IS (261)	261→261	–10				

[a]Diagnostic Ion.

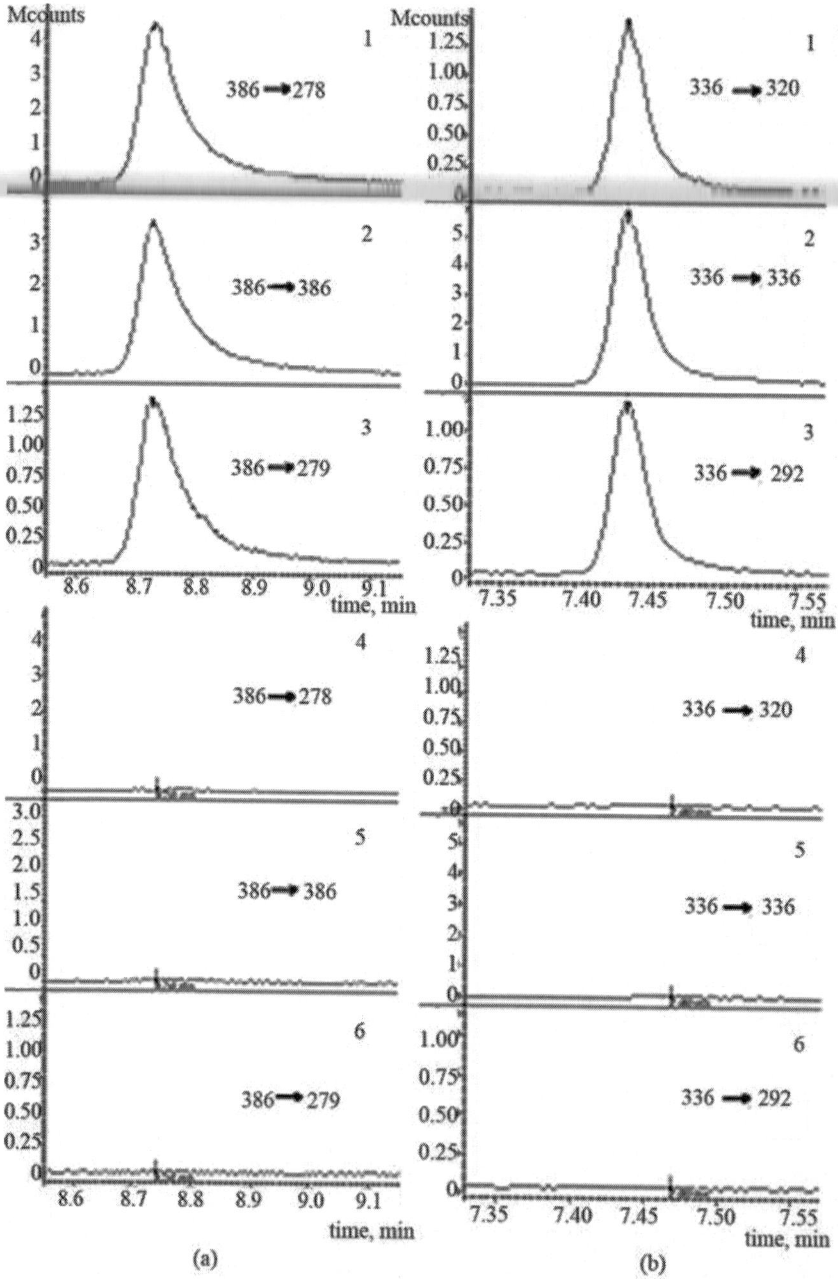

Figure 3. The chromatograms of spiked (1-3) and negative urine sample (4-6), (a) Bendroflumethiazide; (b) Triamterene. C = 0.25 μg.ml^{-1}.

Table 4. The maximum tolerance windows[a] for relative ion intensities.

Relative abundance[b] (%)	Tolerance windows
>50%	±10% (absolute)
25% to 50%	±20% (relative)
5% to <25%	±5% (absolute)
<5%	±50% (relative)

[a]Tolerance window—range for relative ion intensities to ensure appropriate uncertainty in identification; [b]Relative abundance—the abundance of a particular ion relative to the most abundant ion monitored expressed as a percentage.

Peak of small intensity with signal-to-noise ratio ≥ 3 is observed for probenecid in chromatogram of negative urine sample (m/z 270→135, **Table 3**). This fact is possibly caused by the influence of matrix. But the integrated intensity of this peak is negligible which according to WADA recommendations [2] can not be a cause of false identification. Besides, there are no peaks with signal-to-noise ratio ≥ 3 in the chromatograms of negative urine sample obtained using other diagnostic ions.

The ratio of intensities of diagnostic ions of acetazolamide, chlorothiazide, furosemide, chlortha-lidone, bumetanide, hydrochlorothiazide, triamterene and probenecid in the standard solutions and spiked urine samples meets WADA requirements (Tables 3 and 5). In case of indapamide and trichlormethiazide this parameter does not fit the tolerance window. Moreover, significant change of relative intensities dependent on concentration is observed which obviously can be explained by matrix influence at relative intensity. Thereby, this method can not be recommended for confirmation of indapamide and trichlormethiazide in urine.

Limits of detection of acetazolamide, probenecid, furosemide, chlorthalidone, bumetanide, hydrochlorothiazide, triamterene, bendroflumethiazide, metolazone by tandem mass spectrometry are 0.050 $\mu g.ml^{-1}$. Limit of detection of chlorothiazide which is 0.500 $\mu g.ml^{-1}$ does not meet WADA requirements. High limit of detection may be explained by its partial dissociation at the column.

Table 5. Relative abundance of diagnostic ions in Multiple Reaction Monitoring mode, established during the method validation.

Analyte	The relation of ions intensity in spiked urine/standard solution (%)					
	Reaction, m/z	Ion 1	Reaction, m/z	Ion 2	Reaction, m/z	Ion 3
Triamterene	336→336	100/100	336→320	26.8/29.0	336→292	23.9/23.7
Acetazolamide	249→83	100/100	249→108	60.9/62.3	249→249	35.1/35.0
Chlorothiazide	248→141	100/100	248→140	53.5/53.4	248→248	26.7/22.1
Hydrochlorothiazide	353→180	100/100	353→288	37.8/38.0	353→353	7.7/5.1
Trichlormethiazide	399→90	100/100	399→192	73.7/69.8	399→399	49.9/25.4
Bendroflumethiazide	386→278	100/100	386→386	89.1/90.3	386→279	30.5/29.4
Metolazone	392→284	100/100	392→392	34.3/32.4	392→285	18.1/18.2
Indapamide	407→132	100/100	407→161	50.3/59.0	407→407	0.9/0.8
Chlorthalidone	287→255	100/100	287→287	104.8/100	287→220	72.0/71.7
Furosemide	372→81	100/100	372→96	30.4/30.8	372→372	3.0/1.8
Bumetanide	406→318	100/100	406→363	53.4/56.0	406→406	12.0/10.7
Probenecid	270→135	100/100	270→199	25.5/35.0	270→270	4.6/4.7

4. CONCLUSIONS

The developed method proves the suitability of gas chromatography and tandem mass spectrometry for confirmation of presence of 8 diuretics and probenecid in urine. It is selective towards acetazolamide, furosemide, bumetanide, hydrochlorothiazide, bendroflumethiazide, metolazone, probenecid, chlorthalidone and triamterene, since the components of urine do not interfere with the detection of diuretics. The method can not be recommended for determination of indapamide, trichlormethiazide and chlorothiazide.

The reduction of time for sample preparation and sufficient recovery of all diuretics merit the method proposed. Optimization of temperature program for diuretics determination at chromatographic column which wasn't previously used for this purpose allowed to achieve chromatographic separation of all diuretics in short time. Selection of diagnostic fragmentation reactions and optimal collision energies allowed to achieve high selectivity.

Limits of detection of diuretics and probenecid by MS-MS are lower than that regulated by WADA and are 0.050 $\mu g.ml^{-1}$.

The method was validated that proves the possibility of its application in anti-doping laboratories.

REFERENCES

1. WADA, "The world Anti-Doping Code—The 2009 Prohibited List: International Standard," World Anti-Doping Agency, Montreal, 2009.

2. WADA, "Minimum Required Performance Limits for Detection of Prohibited Substances (Technical Document TD2009MRPL)," World Anti-Doping Agency, Montreal, 2009.

3. R. Ventura and J. Segura, "Detection of Diuretic Agents in Doping Control," Journal of Chromatography B, Vol. 687, No. 1, 1996, pp. 127-144.

4. M. Mazzarino, X. de la Torre and F. Botrè, "A Screening Method for the Simultaneous Detection of Glucocorticoids, Diuretics, Stimulants, Anti-Estrogens, Beta-Adrenergic Drugs and Anabolic Steroids in Human Urine by LC-ESI-MS/MS," Analytical and Bioanalytical Chemistry, Vol. 392, No. 4, 2008, pp. 681-698.

5. O. J. Pozo, P. Van Eenoo, K. Deventer and F. T. Delbeke, "Development and Validation of a Qualitative Screening Method for the Detection of Exogenous Anabolic Steroids in Urine by liquid Chromatography-Tandem Mass Spectrometry," Analytical and Bioanalytical Chemistry, Vol. 389, No. 4, 2007, pp. 1209-1224.

6. K. Deventer, O. J. Pozo, P. Van Eenoo and F. T. Delbeke, "Qualitative Detection of Diuretics and Acidic Metabolites of Other Doping Agents in Human Urine by HighPerformance Liquid Chromatography-Tandem Mass Spectrometry. Comparison between Liquid-Liquid Extraction and Direct Injection," Journal of Chromatography A, Vol. 1216, No. 31, 2009, pp. 5819-5827.

7. R. Ventura, M. Roig, N. Monfort, P. Saez, R. Berges and J. Segura, "High-Throughput and Sensitive Screening by Ultra-Performance Liquid Chromatography Tandem Mass Spectrometry of Diuretics and Other Doping Agents," European Journal of Mass Spectrometry, Vol. 14, No. 3, 2008, pp. 191-200.

8. V. Morra, P. Davit and P. Capra, "Fast Gas Chromatographic/Mass Spectrometric Determination of Diuretics and Masking Agents in Human Urine. Development and Validation of a Productive Screening Protocol for Antidoping Analysis," Journal of Chromatography A, Vol. 1135, No. 2, 2006, pp. 219-229.

9. L. Amendola, C. Colamonici, M. Mazzarino and F. Botrè, "Rapid Determination of Diuretics in Human Urine by Gas Chromatography-Mass Spectrometry Following Microwave Assisted Derivatization," Analytica Chimica Acta, Vol. 475, No. 1-2, 2003, pp. 125-136.

10. C. Brunelli, C. Bicchi, A. Di Stilo, A. Salomone and M. Vincenti, "High-Speed Gas Chromatography in Doping Control: Fast-GC and Fast-GC/MS Determination of β- Adrenoceptor Ligands and Diuretics," Journal of Separation Science, Vol. 29, No. 18, 2006, pp. 2765-2771.

11. Y.-L. Tseng, M.-H. Shieh, C.-T. Lin and F.-H. Kuo, "Detection of Diuretics in Urine during Sports Events in Taiwan," Tzu Chi Medical Journal, Vol. 16, No. 2, 2004, pp. 69-77.

12. P. Van Eenoo and F. T. Delbeke, "Criteria in Chromatography and Mass-Spectrometry—A Comparison between Regulations in the Field of Residue and Doping Analysis," Chromatographia, Vol. 59, Suppl. 1, 2004, pp. 39- 44.

13. L. Politi, L. Morini and A. Polettini, "A Direct Screening Procedure for Diuretics in Human Urine by Liquid Chromatography-Tandem Mass Spectrometry with Information Dependent Acquisition," Clinica Chimica Acta, Vol. 386, No. 1-2, 2007, pp. 46-52.

14. C. Goebel, G. Trout and R. Kazlauskas, "Rapid Screening Method for Diuretics in Doping Control Using Automated Solid Phase Extraction and Liqud Chromatography-Electrospray Tandem Mass Spectrometry," Analytica Chimica Acta, Vol. 502, No. 1, 2004, pp. 65-74.

15. D. Thieme, J. Grosse, R. Lang, R. K. Mueller and A. Wahl, "Screening, Confirmation an Quantification of Diuretics in Urine for Doping Control Analysis by High-Performance Liquid Chromatography-Atmospheric Pressure Ionization Tandem Mass Spectrometry," Journal of Chromatography B: Biomedical Sciences and Applications, Vol. 757, No. 1, 2001, pp. 49-57.

16. S. J. Park, H. S. Pyo, Y. J. Kim, M. S. Kim and J. Park, "Systematic Analysis of Diuretic Doping Agents by HPLC Screening and GC/MS Confirmation," Journal of Analytical Toxicology, Vol. 14, No. 2, 1990, pp. 84-90.

17. ISO/IEC 17025, "General Requirements for the Competence of Testing and Calibration Laboratories," 2006.

18. Eurachem Working Group, "A Laboratory Guide to Method Validation and Related Topics," Eurachem, Teddington, 1998.

19. WADA, "The International Standard for Laboratories, Version 6.0," World Anti-Doping Agency, Montreal, 2009.

20. E. Hoffmann and V. Stroobant, "Mass Spectrometry. Principles and Applications," 3rd Edition, WILEY-VCH Verlag GmbH & Co. KGaA, Weinheim, 2007.

21. WADA, "Identification Criteria for Qualitative Assays Incorporating Column Chromatography and Mass Spectrometry (Number TD2010IDCR)," World Anti-Doping Agency, Montreal, 2010.

CHAPTER 8

Diamondoid Characterization in Condensate by Comprehensive Two-Dimensional Gas Chromatography with Time-of-Flight Mass Spectrometry: The Junggar Basin of Northwest China

Shuifu Li [1,*], Shouzhi Hu [1], Jian Cao [2], Ming Wu [2] and Dongmei Zhang [1]

[1]*Key Laboratory of Tectonics and Petroleum Resources (China University of Geosciences), Ministry of Education, Wuhan 430074, China*

[2]*School of Earth Sciences and Engineering, Nanjing University, Nanjing 210093, China*

ABSTRACT

Diamondoids in crude oil are useful for assessing the maturity of oil in high maturation. However, they are very difficult to separate and accurately quantify by conventional geochemical methods due to their low abundance in oil. In this paper, we use comprehensive two-dimensional gas chromatography with time-of-flight mass spectrometry (GC×GC-TOFMS) to study the compounds in condensates from the Junggar Basin of northwest China and address their geological and geochemical applications. GC×GC-TOFMS improves the resolution and separation efficiency of the compounds. It not only separates the compounds that coelute in conventional GC-MS (e.g., 4, 8-dimethyl-diamantane and trimethyl-diamantane) but also allows the identification of compounds that were not previously detected (e.g., trimethyl-diamantane (15A)). A reversed-phase column system improves the separation capabilities over the normal phase column system. The diamondoid indexes indicate that a representative condensate from Well DX 10 is highly mature with equivalent Ro being approximately 1.5%.

KEYWORDS

diamondoid; oil maturity; biomarker; condensate; GC×GC-TOFMS; Junggar Basin

1. INTRODUCTION

Diamondoids are rigid, three-dimensional cyclohexane-ring alkanes that have a diamond-like cage structure [1,2]. Diamondoids in crude oils and source rocks have extremely stable chemical properties. Once formed, they are strongly resistant to thermal degradation and destruction by microorganisms. Therefore, they are generally more stable than other hydrocarbon compounds in geological evolution [3,4]. Because of these factors, diamondoids have important geochemical implications in studying source rocks and oils (especially with high degree of thermal evolution), including maturity, biodegradation and organic facies. This area of research has been a highlight in petroleum geochemistry for many years [3,5–10].

However, the geological and geochemical implications of diamondoids are commonly difficult to address. This is because it is difficult to detect them simultaneously with conventional biomarkers in full scan during conventional GC-MS analysis. In addition, when using selective ion monitoring, they are still not easy to be identified due to their low boiling point and low concentration in oil (some are even in trace amounts). Furthermore, the measurement of diamondoids is conducted on saturated hydrocarbons. Unfortunately, the liquid chromatographic separation of saturated hydrocarbons may result in the loss of the diamondoids due to their low boiling point, which adds to the difficulty in the detection and quantitative analysis of the compounds.

In comparison with conventional GC-MS analysis, the method of comprehensive two-dimensional gas chromatography with time-of-flight mass spectrometry (GC×GC-TOFMS) is effective for separation and identification of complex mixtures, being characterized by high resolution, high sensitivity, large peak capacity, rapid analysis speed and regular qualitative analysis [11–17]. Using this method, diamondoids can be effectively distinguished simultaneously with conventional hydrocarbon compounds [18], providing good conditions for the study of diamondoids. In this paper, we used this method to analyze over 30 condensates collected from the western Junggar Basin of northwest China for the characterization of diamondoids and address their geological and geochemical implications.

2. SAMPLES AND METHODS

The most representative condensate that has the best analytical results was collected from Well DX10 in the western Junggar Basin, with a depth at 3024–3048 m and a Carboniferous age. The condensate has a density of 0.7624 g/cm^3, viscosity of 0.78 mPa.s, freezing point of –24.0 °C and wax content of 0.1%. The condensate was washed in a silica gel column by a solution consisting of dichloromethane and n-hexane in a proportion of 2:1. This aims to collect hydrocarbon fractions, which subsequently were condensed using a nitrogen evaporator and transferred for analysis. The silica gel is 100–200 mesh and was activated at 200 °C for 4 h.

We used a LECO GC×GC-TOFMS instrument system based on an Agilent 7890 N gas chromatograph fitted with a liquid nitrogen-cooled pulse jet modulator and a hot air pulse jet modulator. Each sample was injected in splitless mode. The injection volume was 0.2 μL and the temperature was 310 °C. The experimental system was verified with standard samples, which is prepared with nC_{15}, nC_{16}, nC_{17}, nC_{18} and nC_{19} compounds.

Both normal phase column system and reversed phase column system were conducted. This aims at a comparative study. For the normal system, the inlet temperature was 310 °C, the first-dimension column was a nonpolar HP-5MS (60 m × 0.25 mm × 0.25 μm) that was held at 40 °C for 2 min and then ramped to 310 °C at 2 °C/min, and then maintained for 16 min. The modulator cold jet gas was dry N_2, chilled with liquid N_2. The modulator temperature offset was 30 °C higher than the main GC oven temperature. The modulation time is 6 s with a 1.5 s hot pulse time. Second dimension separations were performed using a DB-17HT (1.2 m × 0.1 mm × 0.1 μm) that was held at 45 °C for 2 min and then ramped to 315 °C at 2 °C/min, and then maintained for 16 min. The carrier gas was He at a constant flow rate of 1.5 mL/min.

For the reversed system, it consists of DB-17 (30 m × 0.25 mm × 0.25 μm) coupled with DB-5 (1.2 m × 0.1 mm × 0.1 μm). The inlet temperature was 300 °C, the first-dimension column was a polar DB-17 (30 m × 0.25 mm × 0.25 μm) that was held at 40 °C for 2 min and then ramped to 300 °C at 2 °C/min, and keep it for 7 min. The modulator cold jet gas was dry N_2, being chilled with liquid N_2. The modulator temperature offset was 30 °C higher than the main GC oven temperature. The modulation time is 7 s with a 1.7 s hot pulse time. Second dimension separations were performed using a DB-5HT (1.2 m × 0.1 mm × 0.1 μm) that was held at 45 °C for 2 min and then ramped to 305 °C at 2 °C/min, and then maintained for 7 min. The carrier gas was He at a constant flow rate of 1.0 mL/min.

The TOF-MS detector signal was sampled at 100 spectra/s and scan range of 55–550 amu. The transfer line held at a constant temperature of 280 °C. The TOF source temperature was 240 °C and the detector was set to −1475 V.

The software of Chromas TOF version 4.43 was used for data processing with database NIST 05. The data processing was mainly conducted by setting up several parameters in the software, such as baseline offset below (usually 0.5), peak width in first dimension (usually it is four times as the modulation time) and peak width in the second dimension (usually approximately 0.1 s) Under such conditions, the software combines fragment peaks whose similarity is over 700, which is an adjustable parameter and usually is 800 in the GC×GC study of oil geochemistry. By this, the software calculates the integrated peak areas automatically. Subsequently, the identification and recognition of compounds are realized by comparing the retention time of the NIST 05 database, combined with a comparison with a professional handbook of chromatograms and mass spectra of known petroleum and geological compounds [19–21].

3. RESULTS AND DISCUSSION

Analytical results show that the diamondoid compounds identified in our work mainly include adamantanes and diamantanes. Their identification was obtained mainly according to comparison with the results of previous studies [2,7,22]. All the compounds elute in one run, demonstrating the advantage of the GC×GC study.

3.1. Identification of Adamantanes

The molecular formula of adamantanes is C_nH_{2n-4} and the carbon number ranges from C_{10} to C_{14}. Adamantanes were identified by characteristic fragment ions of *m/z* 136, 135, 149, 163, 177 and 191 (C_{10} to C_{13}; Figure 1a). In the first-dimensional chromatogram, some adamantanes overlap with chain isoparaffins, such as nC_{11} and No.2 adamantane (Figure 1a). This leads to inaccurate quantitative analysis of adamantanes. In contrast, in GC×GC-TOFMS analysis, adamantanes can be effectively separated, although they have the same or similar first-dimensional retention time (Figure 1b). The carbon number of adamantanes that were identified in the condensate from Well DX 10 ranged from C_{10} to C_{14}.

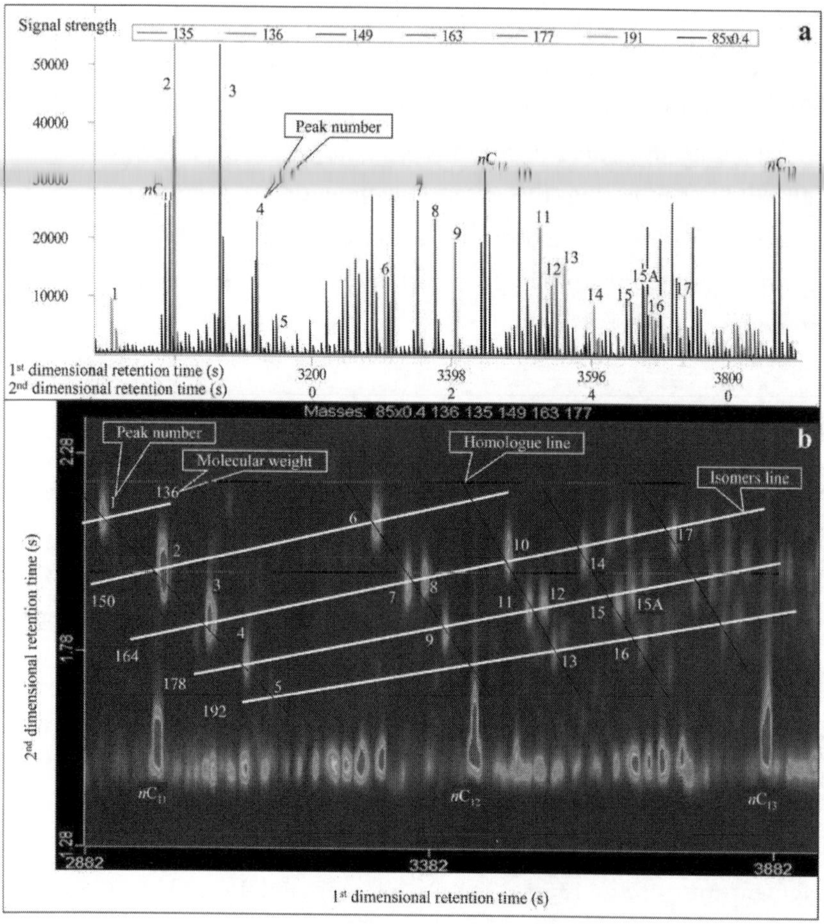

Figure 1. Real second-dimensional chromatogram (**a**) and GC×GC-TOFMS contour plot (**b**) of adamantanes identified in the condensate from Well DX 10 by GC×GC-TOFMS analysis under normal phase column system. The compounds are listed in Table 1.

Table 1. Diamondoid compounds (adamantanes and diamantanes) identified in the condensate from Well DX 10 by GC×GC-TOFMS analysis.

Peak No.	Compound abbreviation	*m/z*	Molecular Weight	First-dimensional retention time (s)	Second-dimensional retention time (s)	Molecular Formula	Full name of compound
1	Adamantane (A)	136	136	2906 (1906*)	2.120 (2.940)	$C_{10}H_{16}$	Adamantane
2	1-MA	135	150	2996 (1878)	1.970 (3.230)	$C_{11}H_{18}$	1-Methyl-adamantane
3	1,3-DMA	149	164	3062 (1843)	1.860 (3.540)	$C_{12}H_{20}$	1,3-Dimethyl-adamantane
4	1,3,5-TMA	163	178	3116 (1801)	1.760 (3.870)	$C_{13}H_{22}$	1,3,5-Trimethyl-adamantane
5	1,3,5,7-TeTMA	177	192	3158 (1745)	1.650 (4.220)	$C_{14}H_{24}$	1,3,5,7-Tetramethyl-adamantane
6	2-MA	135	150	3302 (2249)	2.100 (3.050)	$C_{11}H_{18}$	2-Methyl-adamantane
7	1,4-DMA(*Z*)	149	164	3350 (2214)	1.950 (3.330)	$C_{12}H_{20}$	Cis-1,4-dimethyl-adamantane
8	1,4-DMA(*E*)	149	164	3374 (2179)	1.970 (3.360)	$C_{12}H_{20}$	Trans-1,4-dimethyl-adamantane
9	1,3,6-TMA	163	178	3404 (2130)	1.830 (3.680)	$C_{13}H_{22}$	1,3,6-Trimethyl-adamantane
10	1,2-DMA	149	164	3494 (2361)	2.030 (3.250)	$C_{12}H_{20}$	1,2-Dimethyl-adamantane
11	1,3,4-TMA(*Z*)	163	178	3524 (2305)	1.890 (3.570)	$C_{13}H_{22}$	Cis-1,3,4-Trimethyl-adamantane
12	1,3,4-TMA(*E*)	163	178	3548 (2277)	1.900 (3.600)	$C_{13}H_{22}$	Trans-1,3,4-Trimethyl-adamantane
13	1,2,5,7-TeTMA	177	192	3560 (2221)	1.790 (3.910)	$C_{14}H_{24}$	1,2,5,7-Tetramethyl-adamantane
14	1-EA	135	164	3602 (2459)	2.030 (3.280)	$C_{12}H_{20}$	1-Ethyl-adamantane
15	1-M-3-EA	149	178	3656 (2417)	1.900 (3.530)	$C_{13}H_{22}$	1-Methyl-3-ethyl-adamantane
16	1-E-3,5-DMA	163	192	3686 (2354)	1.800 (3.880)	$C_{13}H_{22}$	1-Methyl-3,5-dimethyl-adamantane
17	2-EA	135	164	3734 (2613)	2.080 (3.220)	$C_{12}H_{20}$	2-Ethyl-adamantane
18	Diamantane (D)	188	188	4724 (3782)	2.539 (2.870)	$C_{14}H_{20}$	Diamantane
19	4-MD	187	202	4766 (3719)	2.370 (3.110)	$C_{15}H_{22}$	4-Methyl-diamantane
20	4,9-DMD	201	216	4808 (3663)	2.220 (3.358)	$C_{16}H_{24}$	4,9-Dimethyl-diamantane
21	1-MD	187	202	4910 (3922)	2.491 (2.990)	$C_{15}H_{22}$	1-Methyl-diamantane

In the trimethyl-adamantanes, there is a compound that has rarely been detected in conventional GC-MS analysis, which is located near the No. 15 peak (Figure 1a). It cannot be accurately named due to the lack of standard spectra and material so far. Since its chromatogram is very similar to the No. 15 peak (Figure 1a), we tentatively named it as the No 15A peak. Its mass spectra is very similar to 1, 3, 4-trimethyl-adamantanes, with *m/z* 163 being the characteristic peak and molecular weight being 178 (Figure 2). Thus, it most likely belongs to the trimethyl-adamantane group of compounds. Furthermore, according to the peak order of multi-methyl-naphthalene and multi-methyl-phenanthrene, the peak of the compound in isomers shows a backward trend when the replacing position of multi-methyl becomes centralized. Therefore, we propose that the compound may be 1, 2, 3- or 2, 3, 4-trimethyladamantanes.

3.2. Identification of Diamantanes

Nine diamantane homologues were identified, ranging from C_{15} to C_{17} (No. 18 to 26 peaks in Figure 3a and Table 1), of which the No. 22 compound is the overlapped peak of 1,2-dimethyl-diamantane and 2,4-dimethyl diamantane. The No. 22, 23 and 24 peaks commonly coelute in conventional GC-MS analysis [18]. In contrast, in GC×GC-TOFMS analysis, the No. 23 and 24 peaks have the same first-dimensional retention time of 4934 s (Table 1). However, their second-dimension retention times are 2.164 s and 2.330 s, respectively, showing a difference of 0.166 s (Table 1). Thus, they can be preliminarily separated based on masses by first-dimensional analysis, and can be effectively separated by second-dimensional analysis (Table 1, Figure 3).

Figure 2. Mass spectra of 1,3,4-trimethyl-adamantane and No. 15A peak in Figure 1a.

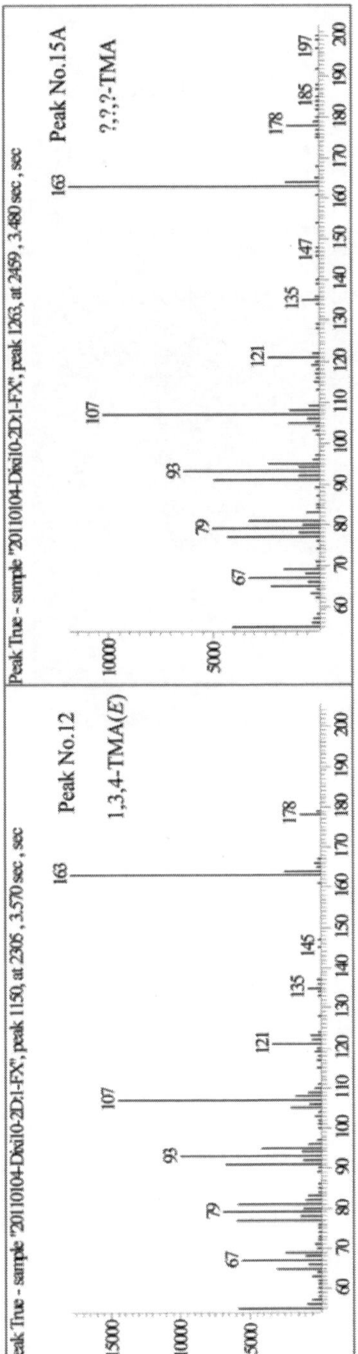

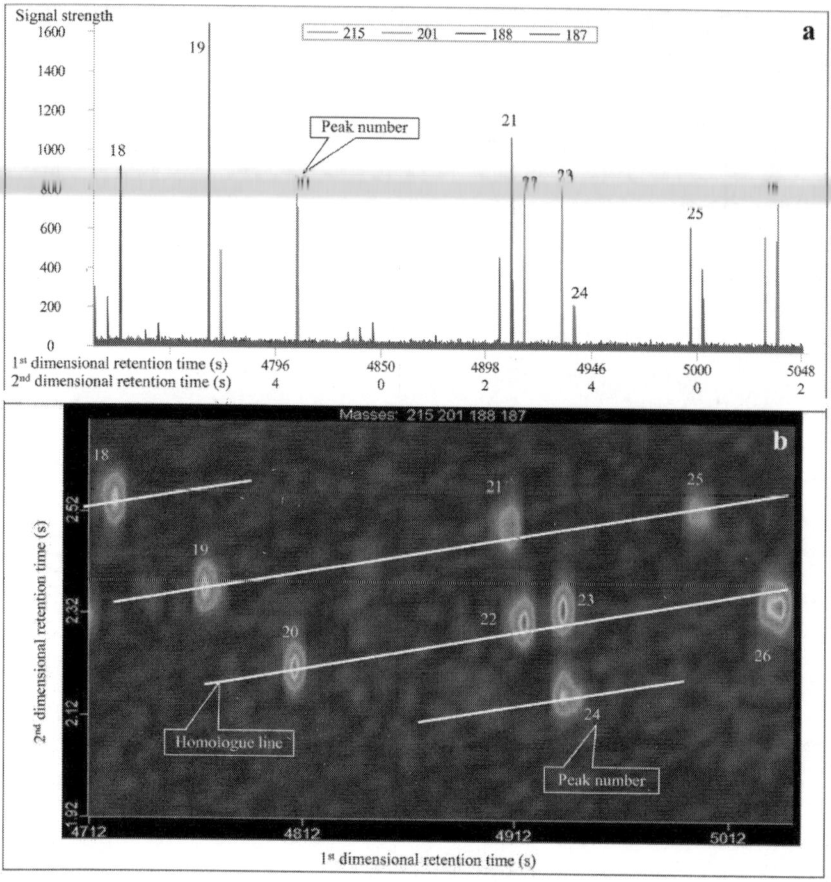

Figure 3. Real second-dimensional chromatogram (**a**) and GC×GC-TOFMS contour plot (**b**) of diamantanes identified in the condensate from Well DX 10 by GC×GC-TOFMS analysis under normal phase column system. The compounds are listed in Table 1.

3.3. Difference between Normal and Reversed Phase Column Systems

Analytical results conducted in normal and reversed phase column systems show differences (Table 1). In the normal system, the first-dimensional and second-dimensional retention times of the compounds with the same characteristic fragment ions increase simultaneously, showing a right-declined imbricate distribution (Figures 1 and 3). By contrast, in the reversed system, the second-dimensional retention time of the compounds decreases while the first-dimensional retention time increases, demonstrating a left-declined imbricate distribution (Figures 4 and 5). In the same second-dimensional chromatogram, the dispersion degree of compounds under the reversed system is obviously larger than that in the normal system, such as the No. 1 and 5 peaks in Table 1, whose distribution span is the largest. Although the second-dimensional retention time interval is only 0.47 s in the normal system, it is far greater in the reversed system (up to 1.28 s; Table 1,Figure 4). With respect to the second-dimensional retention time of the two pairs of cis trans isomers (No. 7 and No. 8, and No. 11 and No. 12), the difference in the reversed system is greater than that in the normal system (Table 1). As to the diamantanes, the first-dimensional retention time interval between No. 18 and 24 peaks is 210 s in the normal system, while it is 245 s in the reversed system (Table 1). In contrast, the second-dimensional retention time interval is 0.375 s and 0.480 s for the normal and reversed systems, respectively. Therefore, in summary, the identification of diamondoids is better in the reversed system (Figures 4 and 5) than in the normal system (Figures 1 and 3).

In our study, the normal and reversed column systems have similar experimental conditions in temperatures, pressure/flow of the carrier gas, *etc.* In contrast, the column length of the reversed system (30 m) is lower than that of the normal system (60 m). Nevertheless, the separation efficiency of the reversed system is still better than that of the normal system.

3.4. Geological and Geochemical Applications

As thermal stability of diamondoids varies with different positions of substituents, they can be used to constrain the degree of thermal evolution of source rocks and crude oils [2,6]. In particular, this is effective for oils with high maturation, which is commonly very hard to constrain in oil geological and geochemical studies [23]. In general, due to geochemical structures and natures, 1-methyl-adamantane (1-MA) is more stable than 2-methyl-adamantane (2-MA), and 4-methyl-diamantane (4-MD) is more stable than 1-methyl-diamantane (1-MD) and 3-methyl-diamantane (3-MD) [2]. Thus, the indexes I (1-MA/(1-MA + 2-MA)) andII (4-MD/(1-MD + 3-MD + 4-MD)) that are calculated based on peak area of the compounds increase with maturity, showing good correlations with R_o [2]. These two index values of the condensate from Well DX 10 based on analysis by normal and reversed phase column systems are 80.18% and 3%–45%, respectively, equivalent to an R_o of approximately 1.5%. This is difficult to determine by conventional geochemical analyses in previous studies. As stated previously, to quantitatively constrain the maturity of oil during its relatively high maturation stage is very difficult. Those commonly-used biomarker parameters that are effective in studying oil maturity under low to moderate maturation stage will have irregular and complex variations, e.g., C_{29} sterane 20S/(20S+20R), C_{29} sterane ββ/(ββ+αα) and Ts/(Ts+Tm) [23]. Thus, there are few studies on the quantitative characterization of the maturity of the DX 10 condensate.

4. CONCLUSIONS

1. Diamondoids can be effectively identified and characterized by GC×GC-TOFMS analysis, which is difficult to achieve using conventional GC-MS measurements. New trimethyl-adamantane 15A (the No. 15A peak in Table 1) compound was detected, which is present close to 1-methyl-3-ethyl-adamantane (the No. 15 peak in Table 1). It may be 1,2,3- or 2,3,4-trimethyladamantanes.

2. Diamondoids that cannot be effectively separated in conventional GC-MS measurements can be effectively separated in GC×GC-TOFMS analysis, such as 4, 8-dimethyl-diamantane (No. 23 peak in Table 1) and trimethyl-diamantane (No. 24 peak in Table 1). This leads to more accurate quantitative analysis and calculations.

3. The separation degree of diamondoids is affected by multiple factors and is generally better in a reversed phase column system than in a normal system.

4. According to diamondoid indexes indicative of maturity, the condensate from Well DX 10 is highly mature with equivalent R_o being approximately 1.5%, obtaining a new effective quantitative indicator. Thus, we obtained new understanding by GC×GC-TOFMS study on diamondoids, whose prospects in the study of petroleum geochemistry were shown.

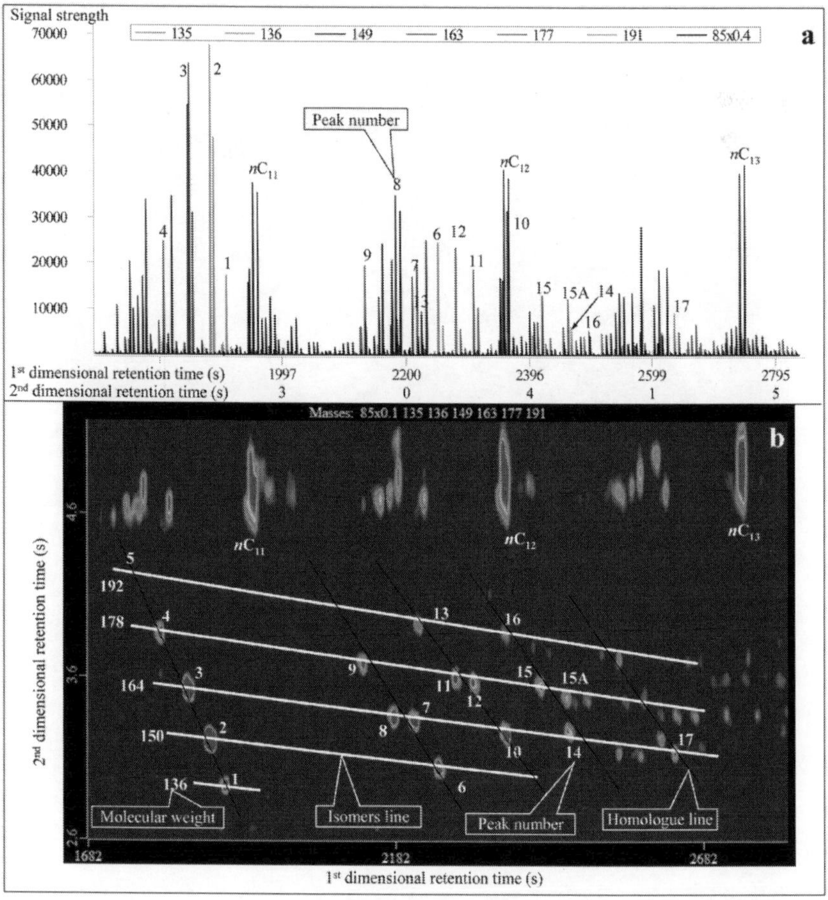

Figure 4. Real second-dimensional chromatogram (**a**) and GC×GC-TOFMS contour plot (**b**) of adamantanes identified in the condensate from Well DX 10 by GC×GC-TOFMS analysis under reversed phase column system. The compounds are listed in Table 1.

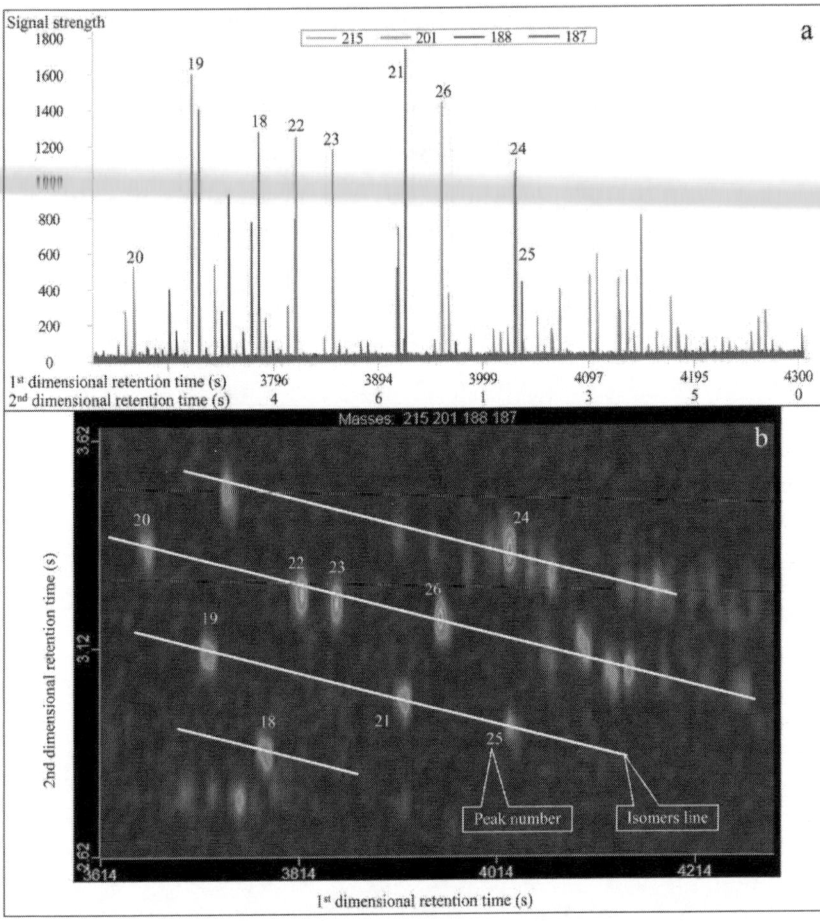

Figure 5. Real second-dimensional chromatogram (**a**) and GC×GC-TOFMS contour plot (**b**) of diamantanes identified in the condensate from Well DX 10 by GC×GC-TOFMS analysis under reversed phase column system. The compounds are listed in Table 1.

ACKNOWLEDGEMENTS

We would like to thank four anonymous reviewers and Chelly Cheng who is the editor of this article. Their detailed and constructive reviews substantially improved the original article. This study was jointly supported by the National Natural Science Foundation of China (Grant No. 41273052, No. 40902037), Fundamental Research Funds for the Central Universities (Grant No. CUGL090219) and the Open Project

Program of Key Laboratory of Tectonics and Petroleum Resources (China University of Geosciences), Ministry of Education (Grant No. TPR-2010-14).

REFERENCES

1. Wingert, W.S. GC-MS analysis of diamondoid hydrocarbons in smackover petroleum. *Fuel* **1992**, *71*, 37–43.

2. Chen, J.H.; Fu, J.M.; Sheng, G.Y.; Liu, D.H.; Zhang, J.J. Diamondoid hydrocarbon ratios: Novel maturity indices for highly mature crude oils. *Org. Geochem* **1996**, *25*, 179–190.

3. Grice, K.; Alexander, R.; Kagi, R.I. Diamondoid hydrocarbon ratios as indicators of biodegradation in Australia crude oils. *Org. Geochem* **2000**, *31*, 715–725.

4. Wei, Z.; Moldowan, J.M.; Paytan, A. Diamondoids and molecular biomarkers generated from modern sediments in the absence and presence of minerals during hydrous pyrolysis. *Org. Geochem* **2006**, *37*, 891–911.

5. Dahl, J.E.; Moldowan, J.M.; Peters, K.E.; Claypool, G.E.; Rooney, M.A.; Michael, G.E.; Mello, M.R.; Kohnen, M.L. Diamondoid hydrocarbons as indicators of oil cracking. *Nature* **1999**, *399*, 54–56.

6. Li, J.G.; Philip, P.R.; Cui, M.Z. Methyl diamantane index (MDI) as a maturity parameter for Lower Palaeozoic carbonate rocks at high maturity and overmaturity. *Org. Geochem* **2000**, *31*, 267–272.

7. Lin, R.; Wilk, Z.A. Natural occurrence of tetramantane ($C_{22}H_{28}$), pentamantane ($C_{26}H_{32}$) and hexamantane ($C_{30}H_{36}$) in a deep petroleum reservoir. *Fuel* **1995**, *74*, 1512–1521.

8. Azevedo, D.A.; Tamanqueira, J.B.; Dias, J.C.M.; Carmo, A.P.B.; Landau, L.; Goncalves, F.T.T. Multivariate statistical analysis of

diamondoid and biomarker data from Brazilian basin oil samples. *Fuel* **2008**, *87*, 2122–2130.

9. Jalees, M.I.; Bianchi, T.S.; Sassen, R.; Tahira, F. Diamondoids and biomarkers; as a tool to better define the effects of thermal cracking and microbial oxidation on oils/condensates from reservoirs of the Upper Indus Basin, Pakistan. Carbonate. *Evaporite* **2011**, *26*, 155–165.

10. Schulz, L.K.; Wilhelms, A.; Rein, E.; Steen, A.S. Application of diamondoids to distinguish source rock facies. *Org. Geochem* **2001**, *32*, 365–375.

11. Eiserbeck, C.; Nelson, R.; Grice, K.; Curiale, J.; Reddy, C. Comparison of GC-MS, GC-MRM-MS, and GC×GC to characterise higher plant biomarkers in Tertiary oils and rock extracts. *Geochim. Cosmochim. Acta* **2012**, *87*, 299–322.

12. Ventura, G.T.; Simoneit, B.R.T.; Nelson, R.K.; Reddy, C. The composition, origin and fate of complex mixtures in the maltene fractions of hydrothermal petroleum assessed by comprehensive two-dimensional gas chromatography. *Org. Geochem* **2012**, *45*, 48–65.

13. Aguiar, A.; Silva, A.I.; Azevedo, D.A.; Aquino, F.R. Application of comprehensive two-dimensional gas chromatography coupled to time-of-flight mass spectrometry to biomarker characterization in Brazilian oils. *Fuel***2010**, *89*, 2760–2768.

14. Ventura, G.T.; Raghuraman, B.; Nelson, R.K.; Mullins, O.C.; Reddy, C.M. Compound class oil fingerprinting techniques using comprehensive two-dimensional gas chromatography (GC×GC). *Org. Geochem* **2010**, *41*, 1026–1035.

15. Li, S.F.; Hu, S.Z.; He, S.; Ma, J.; Liu, G.Q.; Hou, Y.G. Analysis of common compounds in crude oils with two-dimensional gas chromatography time-of-flight mass spectrometry (GCXGC-TOFMS) (in Chinese with English abstract). *Geol. Sci. Technol. Inf* **2010**, *29*, 46–50.

16. Eiserbeck, C.; Nelson, R.K.; Grice, K.; Curiale, J.; Reddy, C.M.; Raiteri, P. Separation of 18α(H)-, 18β(H)-oleanane and lupane by comprehensive two-dimensional gas chromatography. *J. Chromatogr. A* **2011**, *1218*, 5549–5553.

17. Silva, R.S.F.; Aguiar, H.G.M.; Rangel, M.D.; Azevedo, D.A.; Aquino Neto, F.R. Comprehensive two-dimensional gas chromatography with time of flight mass spectrometry applied to biomarker analysis of oils form Colombia. *Fuel***2011**, *90*, 2694–2699.

18. Wang, H.T.; Weng, N.; Zhang, S.C.; Chen, J.P.; Wei, C.Y. Characteristics and identification of saturated hydrocarbons by comprehensive two-dimensional gas chromatography coupled to time-of-flight mass spectrometry (in Chinese with English abstract). *J. Chin. Mass Spectrom. Soc* **2010**, *31*, 18–27.

19. Philip, R.P. *The Spectrum and Application of Fossil Fuel Biomarkers (in Chinese)*; Science Press: Beijing, China, 1987.

20. Zeng, X.Z.; Jiang, J.P.; Jiang, N.H.; Chen, Z.L.; Zhou, H. The criterion of petroleum and natural gas industry, China. In*The Handbook of Mass Spectrum Criterion of Petroleum Biomarkers SY5397–91 (in Chinese)*; Petroleum Industry Press: Beijing, China, 1992.

21. Wang, P.R. *The Mass Spectrum Criterion of Petroleum Biomarkers (in Chinese)*; Petroleum Industry Press: Beijing, China, 1993.

22. Berwick, L.; Alexander, R.; Pierce, K. Formation and reactions of alkyl adamantanes in sediments: Carbon surface reactions. *Org. Geochem* **2011**, *42*, 752–761.

23. Peters, K.E.; Walters, C.C.; Moldowan, J.M. *The Biomarker Guide, Biomarkers and Isotopes in Petroleum Exploration and Earth History*; Cambridge University Press: Cambridge, UK; p. 2005.

CHAPTER 9

Optimization of Solid Phase Microextraction Procedure Followed by Gas Chromatography with Electron Capture Detector for Pesticides Butachlor and Chlorpyrifos

Fateme Ghavidel[1], Seyed Jamaleddin Shahtaheri[1,2*], Reza Khani Jazani[3], Marzieh Torabbeigi[3], Abbas Rahimi Froushani[4], Monireh Khadem[1]

[1]*Department of Occupational Health Engineering, School of Public Health, Tehran University of Medical Sciences, Tehran, Iran*

[2]*Institute for Environmental Research, Tehran, Iran*

[3]*Faculty of Health, Safety, and Environment, Shahid Beheshti University of Medical Sciences, Tehran, Iran*

[4]*Department of Epidemiology and Biostatistics, School of Public Health, Tehran University of Medical Sciences, Tehran, Iran*

ABSTRACT

In this study, headspace solid phase microextraction (HS-SPME) followed by gas chromatography using electron capture detection system (GC-ECD) were developed for the determination of chloraacetanilide (butachlor) and chlorpyrifos presented in biological samples. Different parameters affecting the extraction procedure were optimized including extraction time (30 minutes), extraction temperature (80°C), sample volume (3 mL), sample pH

(2), added NaCl (0.3 gram) and sample stirring rate (400 rpm). Different concentrations of 1 - 100 ng/ml were applied for butachlor and a linear calibration curve was obtained. Furthermore, a similar linearity was obtained for chlorpyrifos, using a concentration range of 1 - 250 ng/ml. The limit of detection (LOD) obtained for butachlor and chlorpyrifos were 0.000 and 0.53 ng/ml respectively. The optimized methods for both compounds were validated using two concentrations of 25 and 50 ng/ml in spiked urine samples. Obtained recoveries of spiked urine samples were 83.06% - 99.8% with RSD of lower than 11%. Optimized technique was simple, inexpensive, solvent free and fast in comparison with other conventional methods and had compatibility with the chromatographic analytical system. This method offers low detection limits to analyze pesticides in urine samples that are very important in the exposure monitoring in occupational health.

KEYWORDS

HS-SPME, GC-ECD, Biological Samples, Chlorpyrifos, Butachlor

1. INTRODUCTION

Pesticides are widely used for agricultural purposes around the world. Application of pesticides in agriculture has increased productivity and has led to the increase yields in farming. Pesticides are biologically active and can be harmful to the variety of non-target organisms (including human) when they are applied [1] [2]. Therefore, nowadays, toxic effects of pesticides on humans and the environment are of very important concern [3]. Among the various pesticides available; butachlor and chlorpyrifos are the ones most widely used in developing countries, including Iran by the relevant plant protection organization. Butachlor is structurally belonging to the chloroacetanilide family, and chlorpyrifos is in the organophosphorus pesticide category. It is worth mentioning that, butachlor is a suspected carcinogen [4], so, it needs to be more monitored in the environment, and as a result, monitoring this pesticide in the environment and human is most important for evaluation of the risk hazards associated with such pesticides [5] . Chlorpyrifos as organophosphorus pesticides is very toxic when absorbed by human being because of its property of acetyl-cholinesterase inhibition [6] . To evaluate exposure to these pesticides and the corresponding health risk, measuring the parent pesticide levels in body fluids may be quite useful [7] [8] . There are some traditional techniques to determine the pesticides such as liquid chromatography and gas chromatography with very low detection limits. But the classical gas chromatographic methods are not satisfactory due to the thermal instability of the organophosphorus molecules and a sample preparation or pre-concen- tration step is needed to determine the analytes at ppb levels. Several analytical methodologies have been developed for sample preparation to facilitate identification of pesticides from biological and environmental samples. Liquid-liquid extraction (LLE) is one of the conventional methods still being used for the extraction of such pesticides from various matrices [9] [10] . Also, solid-phase extraction (SPE) [11] - [17] , supercritical fluid extraction (SFE) [18] , and matrix solid-phase dispersion (MSPD) [19] is the other more selective approach in this field. Generally, these methods are multistage, expensive, laborious, and consume long time as well as; in these methods, a large amount of organic solvents are applied to extract the analysts of interest [20] .

Modern microextraction techniques such as solid-phase microextraction (SPME), that is a rapid, inexpensive, solvent free and quantitative analytical technique was developed by Pawliszyn et al. in the early 1990s [21] [22] . SPME has been employed as a sample preparation method for the analysis of pesticides in various matrices [23] - [25] . In SPME, headspace (HS) approach is preferred for extracting analytes from biological samples, including blood and urine because of complexity of these matrices [26] . The previous studies about chlorpyrifos have used HS-SPME-GC-ECD and mass method (MS) in water samples, fruits, vegetable, and milk [27] [28] . Moreover, Direct Immersion (DI)-SPME-GC-MS technique has been performed for determining the chlorpyrifos in water samples [29] . HS-SPME procedure has been applied for pesticide's purification in biological samples such as chloraacetanilide [30] and organophosphorus pesticides [31] .

In comparison with other methods, there are several more variables affecting the determination of butachlor and chlorpyrifos by HS-SPME method that must be studied and optimized to obtain high analyte recovery and low detection limits. Under the optimum extraction conditions, high extraction efficiency was achieved in a relatively short time. Also, there are very few existing methods to determine butachlor and chlorpyrifos in urine samples specially for monitoring the workers exposure to these compounds. The aim of the study was to develop a HS-SPME procedure by optimizing several extraction parameters, and to validate the optimized method by performing urine sample analysis for simultaneous determination of butachlor and chlorpyrifos. Parameters affecting the extraction efficiency were optimized, i.e. extraction temperature, extraction time, salt addition, sample pH, sample volume, and stirring rate. Through this study, optimum conditions were examined in the real samples. The validation of spiked samples was carried out using the optimum conditions. Limit of detection, limit quantification, and linearity range of the method was established for evaluating the pesticides. This method can lead to low detection limits to analyze pesticides in urine samples that are very important in the occupational exposure monitoring.

2. MATERIALS AND METHODS

2.1. Reagent and Materials

Standards of pesticides were purchased from Sigma-Aldrich (Germany) including butachlor (97.7%) and chlorpyrifos (99.9%). A stock solution of butachlor (1000 $mg \cdot L^{-1}$) and chlorpyrifos (2500 $mg \cdot L^{-1}$) was prepared in acetone and methanol, respectively and stored at 4°C. All solvents (methanol, acetone, and water) were HPLC grade (Merck-Germany). Other reagents with an analytical grade were obtained from Merck (Germany). Working standard solutions of both pesticides were freshly prepared daily by volume dilution in the HPLC grade water.

2.2. Biological Samples

After obtaining approval from Tehran University of Medical Sciences (Vice-Chancellor for Research), all participants signed related consent form for the voluntary participation. Then their urine samples were collected for optimization studies. It is worth to mention that a limited number of healthy and non-exposed persons were randomly selected to study. Urine samples were collected in mornings under documented consent, prepared by adding the appropriate volume of the methanol standard solution, and stored at 4°C until use. These urine samples were adopted to test the recovery and reproducibility of the optimized method. Since one of the aims of this study was to introduce a proper method for determination of trace level of pesticides in biological samples, so the urine samples were selected to approve the obtained extraction efficiency in this kind of samples. To test the method recovery using the biological samples, defined concentrations of pesticides were spiked in urine (related to non-exposed persons) and optimized extraction procedure was used to get the recoveries.

2.3. HS-SPME Extraction Procedure

The SPME device for manual extraction purchased from Supelco (Bellefonte, PA, USA). Polydimethylsiloxane (PDMS 100 μm) fibres were used as extraction phase because of its great sensitivity. Before using, fibers were conditioned in the injection port of the gas chromatograph for 0.5 h at 250°C according to the recommended instructions by the manufacturer. SPME procedure was performed in biological samples by placing 500 μl of urine sample and appropriate volumes of pesticide solution at two-level concentrations of 25 and 50 ng/ml in 5 volumetric flasks. These levels with defined concentrations were selected as spiked samples to optimize variables of interest. Hydrochloric acid and sodium chloride was added for adjusting the pH at 2 and salt was added to reach the concentration of 10% and then HPLC grade water was used for adjusting. Three ml volume of this solution was transferred to a 10 ml vial. The vial was rapidly sealed with a silicon septum and aluminum cap. Then; the vial was heated up to 80°C in aluminum block heater. Samples were stirred during the extraction time under the magnetic stirring (400 rpm). SPME needle pierced the vial septum in order to expose the fiber (100 μm PDMS) on the headspace of the solution. After 30 minutes, the needle was removed from the vial and inserted into the heated injection port of the GC-ECD for thermal desorption of the sample for 5 minutes.

2.4. Gas Chromatography

Chromatographic analysis was carried out by a Varian CP 3800 gas chromatograph equipped with an electron capture detector (ECD). Pesticides were separated in a HP-5 (30 m × 0.32 mm, 0.25 m film thickness) column. Nitrogen (99.999%) with a flow rate of 2 ml/min was used as a make-up and carrier gas. Samples were analyzed using the following conditions: oven temperature program: initial temperature: 80°C hold for 1 min, rate 30°C to 200°C hold for 2 min, rate 10°C to 250°C hold for 3 min, rate 20°C to 280°C hold for 4 min. The injector temperature was 280°C. The injector was operated in the split less mode. Detector temperature (ECD) was set at 300°C.

2.5. Data Analysis

The statistical calculations and analysis were performed using SPSS version 16. Some tests such as pair T-test and scheffe analysis of variance were used for data analyzing. The level of significance was taken as $p < 0.05$. In order to optimize the method, a constant concentration was used for each pesticide. For each level of intended variables, three samples with defined concentration were prepared and the mean of extraction efficiencies calculated after analyzing the analytes. Then according to obtained data, the optimum level was selected for every variable.

3. RESULTS

3.1. Method Optimization

3.1.1. Extraction Temperature

The effect of temperature on extraction efficiency was evaluated from 40°C to 80°C. The highest extraction efficiency for butachlor was obtained at

80°C. This was due to increased transmission of an analyst into the headspace and hence, into the fiber with increasing temperature. The response of chlorpyrifos increased slightly from 40°C up to 60°C. After that, the higher temperature had no significant effect on extraction rate (Figure 1). It means that the response is nearly constant in temperatures from 60°C to 80°C. Since the best response for butachlor obtained in the temperature 80°C, for improving the simultaneous extraction of butachlor and chlorpyrifos, temperature of 80°C was chosen as the optimum temperature.

3.1.2. Extraction Time

The effect of extraction time was evaluated by exposing the fiber to the headspace of the samples at different extraction times, including 5, 10, 30, and 60 minutes. Equilibrium time for butachlor and chlorpyrifos (from 5 minutes to 30 minutes) was remarkable and showed an increased extraction rate. However, larger extraction time for both pesticides was unsuitable, because the responses of both pesticides were decreased when the extraction time exceeded 30 minutes (Figure 2). Therefore, extraction time of 30 minutes was selected at an optimum extraction time, and rms do not have to be defined. Do not use abbreviations in the title or heads unless they are unavoidable.

3.1.3. Salt Addition

The influence of salt on extraction efficiency was investigated for both pesticides by adding various amounts of NaCl in sample solution, including 0%, 1%, 5%, and 10% (w/v) of sodium chloride. As it can be seen in Figure 3, in general, super saturation of the sample with salts is most effective for the extraction of analytes onto the fiber because of the salting-out effect [32] . The response of chlorpyrifos was increased from 0% to 10% of NaCl, and the highest extraction efficiency was obtained in 10% of NaCl. The effect of NaCl on extraction efficiency of butachlor showed that, the amount of NaCl from 0% to 5% increased the efficiency. However, the higher amount of sodium chloride for butachlor showed a slight decrease. Therefore, NaCl 10% was chosen as an optimum amount of salt added to the butachlor and chlorpyrifos samples. Although according to figure, the optimal salt concentration seems to be 5% for both butachlor and chlorpyrifos, there was no significant difference between 5% and 10% of NaCl for both of them. And since the higher ionic strength had no negative impact on the extraction efficiency and even may improve it in presence of other pesticides, 10% of NaCl was selected as proper salt concentration.

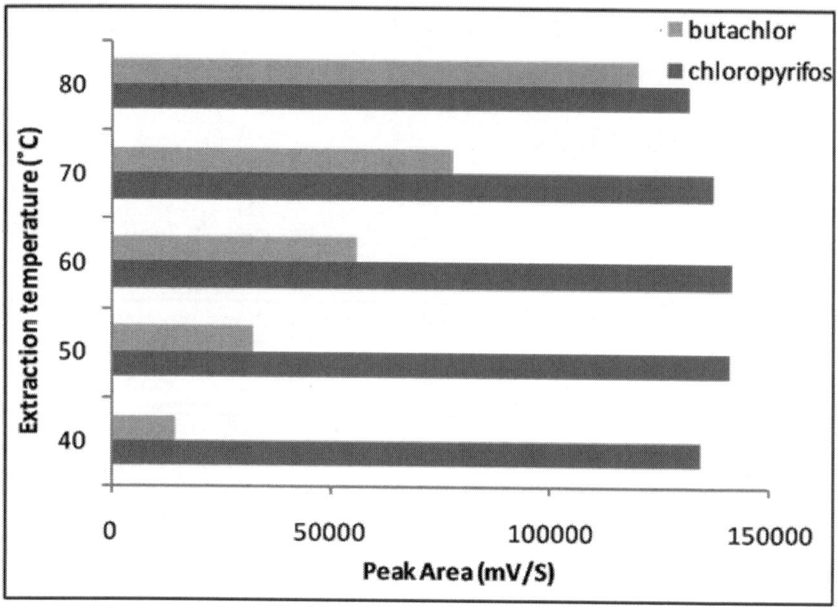

Figure 1. Influence of extraction temperature on the extraction efficiency (Extraction time: 30 min, sample volume: 3 ml, sample pH: 2, NaCl: 10% (0.3 gram), stirring rate: 400 rpm).

3.1.4. Effect of pH

The effects of pH on extraction efficiency for butachlor and chlorpyrifos have been shown in Figure 4. The extraction efficiency of these pesticides was studied at various pHs from 2 to 5. The efficiency of both pesticides was decreased significantly when pH increasing. Hence, a sample pH of 2 was chosen as the optimum pH for each two pesticides.

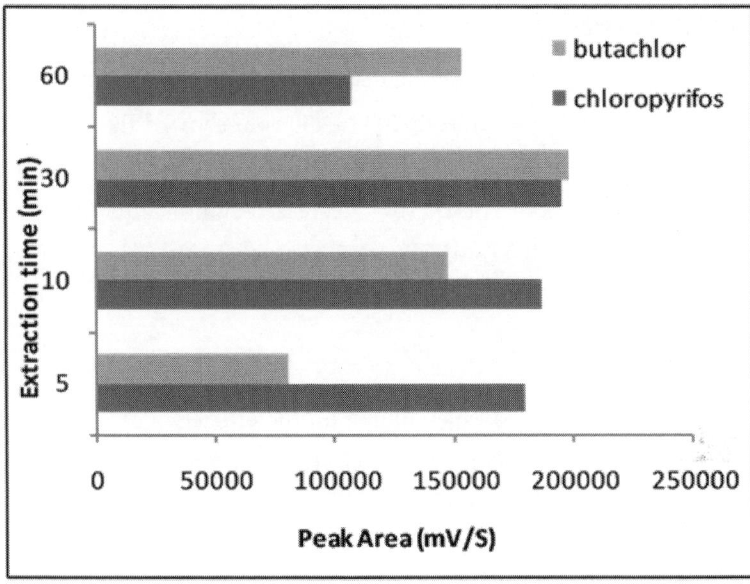

Figure 2. Influence of extraction time on the extraction efficiency (Extraction temperature: 80˚C, sample volume: 3 ml, sample pH: 2, NaCl: 10% (0.3 gram), stirring rate: 400 rpm).

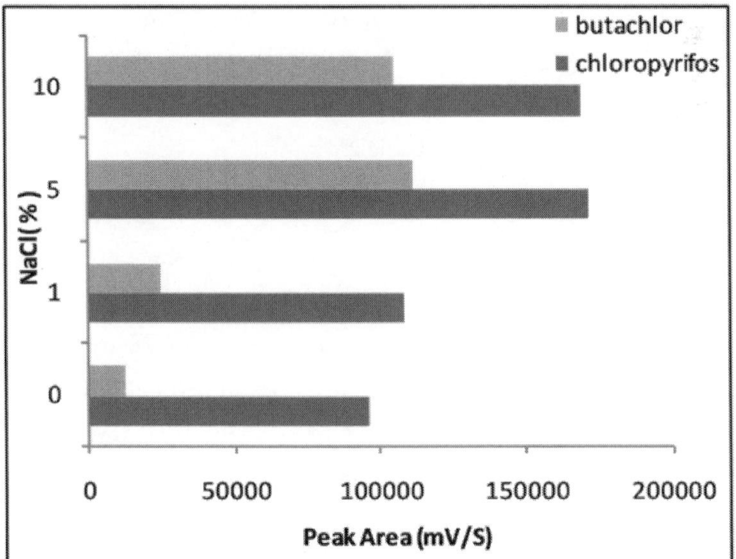

Figure 3. Influence of salt addition on the extraction efficiency (Extraction time: 30 min, extraction temperature: 80˚C, sample volume: 3 ml, sample pH: 2, stirring rate: 400 rpm).

3.1.5. Sample Stirring

The effect of sample stirring was evaluated at different stirring rates (200, 400, and 700 rpm). Stirring rates from 200 to 400 rpm showed that the peak areas of both pesticides were increased with stirring rate. The extraction efficiencies of Butachlor and chlorpyrifos decreased significantly at higher stirring rates (from 400 to 700 rpm) (Figure 5). Therefore, the velocity of 400 rpm was chosen for subsequent analyses.

3.1.6. Sample Volume

To evaluate the effect of the sample volume on the efficiency of HS-SPME, the sample volumes of 2 to 4 ml were selected. For both pesticides, the maximum response was obtained when 3 ml sample was used. At the higher sample volumes, as it can be seen in Figure 6, extraction efficiency showed a considerable decrease. However, subsequent analyses were performed at 3 ml sample volume.

3.2. Method Validation

The method was validated using obtained optimized conditions. The detection limits (LODs), quantification limits (LOQs), the inter- and intra-day precisions (RSD), the linearity, and the recoveries were calculated. The results were tabulated in Table 1 and Table 2.

Linear standard curves in the range of 1 - 100 ng/ml and 1 - 250 ng/ml for butachlor and chlorpyrifos were obtained each day for 6 consecutive days, respectively. The obtained correlation coefficients (r^2) were 0.9928 or greater (Table 1). Intra- and inter-day precisions of the method were determined by two spiked extracted samples at concentrations of 25 and 50 ng/ml followed by analysis each day for 6 consecutive days to estimate inter-day (day-to-day) reproducibility as well as six times a day to estimate intra-day (within the day) reproducibility. Totally, 72 analyses were performed to determine intra and inter reproducibility. In within day experiments evaluation, one urine samples per day were analyzed six times for three consecutive days (36 analyses at two intended concentrations). In day-to-day reproducibility, experiments were performed every day for 6 consecutive days with three replications (36 analyses at two intended concentrations). Figure 7 shows the chromatograms of urine sample with and without spiked pesticides. The optimized extraction protocol was followed along with optimum chromatographic conditions. The measurement of butachlor and chlorpyrifos on each day of the analysis was calculated by comparing it with the extracted calibration curve generated on that day. Intra-day variation or reproducibility of the method was assessed by calculating the relative standard deviation (RSD) (Table 2). As indicated by RSD, the precision of the optimized method was high enough (RSD less than 11%). The relative standard deviation of 50 ng/ml showed lower RSD (higher precision). Experiments were also undertaken to assess Inter-day reproducibility. The results are shown in Table 2. The obtained RSD indicates that the method was much more precise at two concentrations. The recoveries obtained by SPME-GC-ECD method was between 83% - 99% that shown in Table 2.

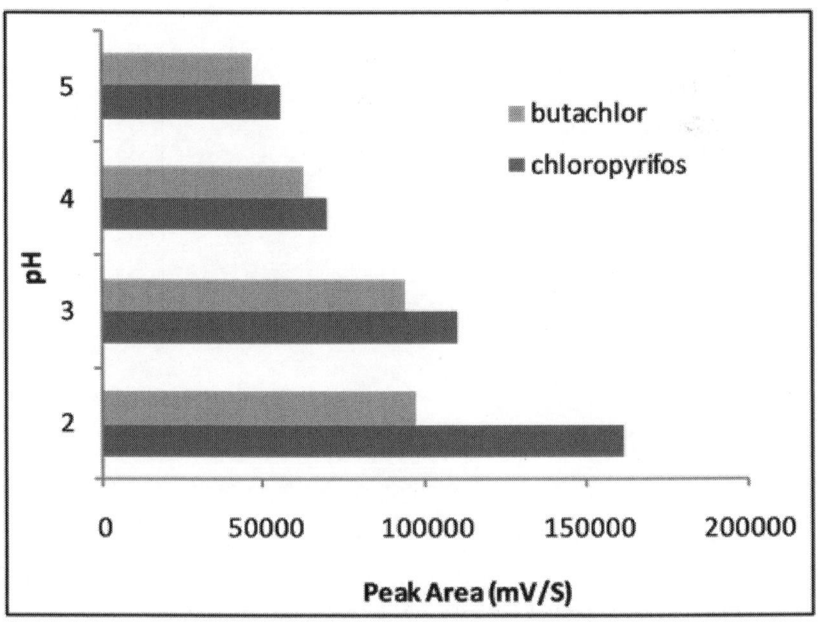

Figure 4. Influence of sample pH on the extraction efficiency chloroprifos and butachlor (Extraction time: 30 min, extraction temperature: 80°C, sample volume: 3 ml, NaCl: 10% (0.3 gram), stirring rate: 400 rpm).

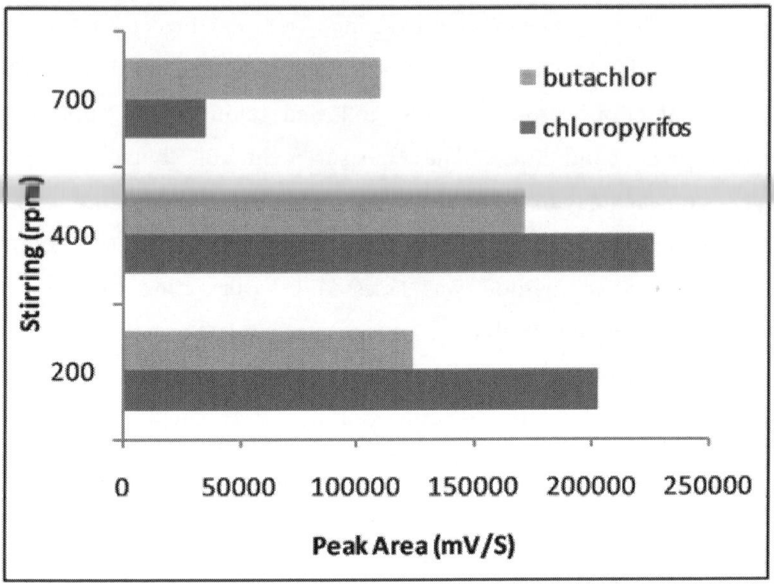

Figure 5. Influence of stirring speed on the extraction efficiency (Extraction time: 30 min, extraction temperature: 80°C, sample volume: 3 ml, sample pH: 2, NaCl: 10% (0.3 gram)).

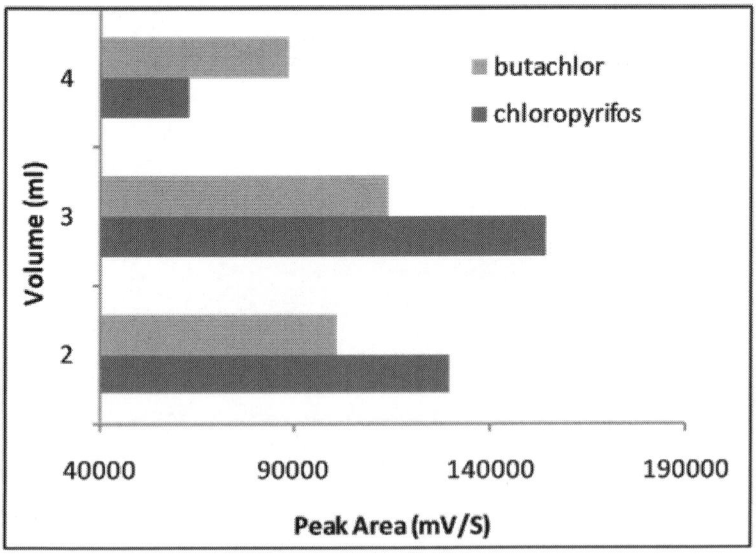

Figure 6. Influence of sample volume on the extraction efficiency (Extraction time: 30 min, extraction temperature: 80°C, sample pH: 2, stirring rate: 400 rpm, NaCl: 10% (0.3 gram)).

Table 1. Analytical performance of the optimized SPME method for two pesticides.

Analyte	Linear rang (ng/ml)	Correlation coefficient (R^2) (ng/ml)	LODs (ng/ml)	LOQs (ng/ml)
Butachlor	1 - 100	0.992	0.0888	0.269
Chlorpyrifos	1 - 250	0.997	0.053	0.161

Table 2. Inter-day and intra-day reproducibility of method for spiked urine sample.

Analytes	Concentration	Inter-day (n = 6)		Intra-day (n = 6)	
		Recovery (%)	RSD (%)	Recovery (%)	RSD (%)
Butachlor	25	90.65	8.098	87.70	8.266
	50	96.70	6.594	97.47	4.601
Chlorpyrifos	25	83.06	10.06	85.59	10.206
	50	99.60	4.833	99.81	5.523

4. DISCUSSION

In this study, headspace solid phase microextraction (HS-SPME) method in combination with Gas Chroma-Tography-Electron Capture Detector (GC-ECD) was used for the extraction and quantification of butachlor and chlorpyrifos. This technique is simple, inexpensive, and fast compared to other conventional methods [21] [22] [33] [34] . Through this study, different parameters affecting the extraction procedure were optimized. Temperature is the main parameter that can affect the extraction of the analyte in HS-SPME. By raising temperature, the partition constant of the analyte was increased in favour of the gaseous sample; therefore, partitioning of the analytes between the sample and headspace can reach equilibrium more quickly. Thus, the analyte can exist in the headspace of the sample and can be extracted by the SPME fiber. Increasing temperature causes an increase in the amount of the analyte extracted. The obtained results indicated that, temperature up to 60°C could increase the amount of the analyte extracted for chlorpyrifos, where it reached a plateau and remained constant. No significant effect was seen at higher temperatures ($p > 0.05$)

and highest extraction efficiency for butachlor was obtained at 80°C. These results are consistent with previously reported data [30] [35].

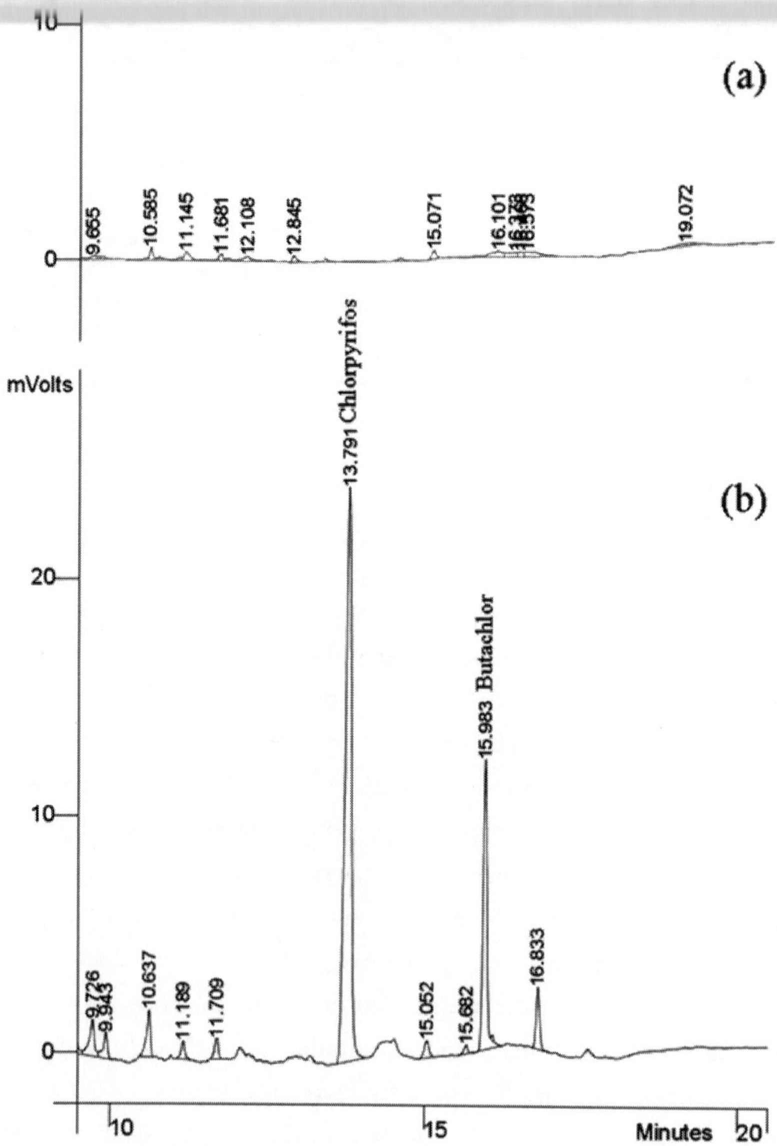

Figure 7. Chromatograms of urine sample (a) without and (b) with spiked sample of chloroprifos and butachlor.

The effect of extraction time of 5 - 60 minutes on extraction efficiency indicated that, up to 30 minutes there was a significant effect in all experiments for both pesticides. It means that based on the peak obtained from the GC, the higher amounts of analytes could efficiently extract from the samples with increasing the time up to 30 minutes. After this time, analytes were desorbed from the fiber to the headspace and the amounts adsorbed on the fiber were slightly decreased. The more extraction time can decrease the distribution coefficient of the analyte [36] .

NaCl is also a parameter that can affect extraction efficiency, because the partition constant of the analyte can be changed in its presence, and therefore, more analyte may be exist in the headspace of the sample. In fact, the solubility of the analyte decreases when NaCl is added.

Because of the salting-out effect, super saturation of the sample with salts can cause the extraction of analytes onto the fiber [32] . Also, previous studies have shown that ionic strength can improve extraction efficiency [37] . Therefore, the effect of extra NaCl on the extraction efficiency of both pesticides was shown. There is a competition among compounds with different polarity for the fiber in SPME. Also intensities of competition in presence of various pesticides are different. So, there are some differences in various studies due to different existing pesticides. For butachlor, the amount of NaCl from 0% to 5% increased the extraction rate; after that, there was no significant difference ($p > 0.05$) between 5% and 10% of NaCl and the extraction remained constant.

The pH of the sample was the main parameter which could theoretically affect extraction efficiency. The ionic strength of the analytes could be changed by increasing or decreasing pH. The extraction efficiency of both pesticides was decreased significantly when pH was increased. In general, the sample is acidified for the extraction of acidic analytes and is made alkaline for the extraction of basic analytes, because, compounds for extraction should be in the molecular form [38] .

Sample agitation is another important parameter that has to be optimized. It can increase extraction efficiency. The stirring process accelerates the transfer of an analyst into the headspace and hence, into the fiber coating. Although the equilibration time decreases with the increasing stirring rate, quick agitation depends on being rotational speed might cause a change in the equilibration time [38] . Higher stirring rate may cause a negative effect on the reproducibility of the sorption condition because of spattering.

For optimizing the volume of the sample, since the size of headspace vials was 10 ml, 2 - 4 ml of the sample volume was tested. By increasing the volume of the sample up to 3 ml, the amount of the analyte extracted can be increased. Gorecki and Pawliszyn in their recent paper have reported that, a large sample volume can provide the best results [39] . At the higher sample volumes, extraction rate showed a considerable decrease due to lower headspace. For rapid extraction, a large capacity of the headspace is required. It is achievable when there is a large headspace-to sample volume ratio. On the other hand, increasing the sample volume in a constant vial size leads to the lower head space capacity. In present study, one vial size was selected because of the poorer precision using variable vial volumes. Therefore, increasing the sample volume decreased the capacity of headspace, resulting lower extraction efficiency. Also the effect of sample volume on the magnitude of extraction depends on the analyte (how much is the analyte soluble and volatile). Gorecki and Pawliszyn have worked on the partition coefficient (K) values on their study and express that the large sample volumes is used to determine K values, as they enable broader ranges of values with good accuracy. On the other hand, they have emphasized that a large headspace capacity may lead to the loss of sensitivity. These reasons may explain the difference between the results of studies.

Finally, validation of the method was performed using spiked samples and standards to determine the pesticides in urine samples. The recoveries and RSD values in the present study were compared with those values in the existing methods [27] [32] [40] and the significant differences were found between them. These mentioned studies expressed the different RSDs and Extraction recoveries (like RSD < 19%), but there were significant differences between them and present study. The obtained recoveries in this study, ranging from 88% to 97%, were higher than the mentioned investigations. The optimized method is sensitive enough for all concentrations (RSD less than 11%). Based on results, the proposed method is easy, fast, precise, and reproducible with wide linear range and it does not require any toxic solvent for extraction. In addition, it can be seen a successful application of the HS-SPME technique to analyze the pesticides at a low concentration level in urine samples.

5. CONCLUSION

The SPME procedure seems to be a useful method for qualitative and quantitative evaluation of butachlor and chlorpyrifos in biological samples. This technique is simple, inexpensive, fast compared to the more conventional methods. To implement HS-SPME-GC-ECD, it was necessary to optimize the parameters affecting the extraction of the analytes. The optimum extraction conditions were achieved. Finally, validation of methods for determining butachlor and chlorpyrifos in the spiked urine sample was performed. Inter-day and intra-day relative standard deviation of method was evaluated by spiked urine sample with butachlor and chlorpyrifos. The results indicated that, the developed SPME-GC-ECD method was suitable for biological monitoring of butachlor and chlorpyrifos.

ACKNOWLEDGEMENTS

This research has been supported by Tehran University of Medical Sciences grant (Project No. 15212). Authors acknowledge the University for all valuable supports.

REFERENCES

1. Hsu, J.P., Herbert, G., Wheeler, D.E., Camann, H.J., et al. (1988) Analytical Methods for Detection of Nonoccupational Exposure to Pesticides. Journal of Chromatographic Science, 26, 181-189. http://dx.doi.org/10.1093/chromsci/26.4.181

2. Matsumura, F. (1985) Toxicology of Insecticides. 2nd Edition, Plenum, New York.

3. Ritter, L., Goushleff, N.C.I., Arbuckle, T., Cole, D., Donald, M. and Raizenne, M. (2006) Addressing the Linkage be- tween Exposure to Pesticides and Human Health Effects—Research Trends and Priorities for Research. Journal of Toxicology and Environmental Health, 9, 441-456.

4. Farombi, O.E., Ajimoko, Y.R. and Adelowo, O.A. (2008) Effect of Butachlor on Antioxidant Enzyme Status and Lipid Peroxidation in Fresh Water African Catfish, (Clarias gariepinus). International Journal of Environmental Research and Public Health, 5, 423-427.

5. Aktar, W., Sengupta, D. and Chowdhury, A. (2009) Impact of Pesticides Use in Agriculture: Their Benefits and Ha- zards. Interdisciplinary Toxicology, 2, 1-12.

6. Costa, L.G. (2006) Current Issues in Organophosphate Toxicology. Clinica Chimica Acta, 366, 1-13.

7. Otero, R., Santiago Silva, M. and Grimalt, J.O. (1997) Hexachlorocyclohexanes in Human Blood Serum. Journal of Chromatography A, 778, 87-94.

8. Burse, V.W., Head, S.L., Korver, M.P., McClure, P.C., Donahue, J.F. and Needham, L.L. (1990) Determination of Se- lected Organochlorine Pesticides and Polychlorinated Biphenyls in Human Serum. Journal of Analytical Toxicology, 14, 137-142.

9. Pico, Y., Font, G., Molto, J.C. and Manes, J. (2000) Pesticide Residue Determination in Fruit and Vegetables by Liquid Chromatography-Mass Spectrometry. Journal of Chromatography A, 882, 153-173.

10. Jimenez, J.J., Bernal, J.L., del Nozal, M.J., Novo, M., Higes, M. and Llorente, J. (2000) Determination of Rotenone Residues in Raw Honey by Solid-Phase Extraction and High-Performance Liquid Chromatography. Journal of Chro- matography A, 871, 67-73.

11. Tsydenova, O.V., Sudaryanto, A., Kajiwara, N., Kunisue, T., Batoev, V.B. and Tanabe, S. (2007) Organohalogen Compounds in Human Breast Milk from Republic of Buryatia, Russia. Environmental Pollution, 146, 225-235.

12. Shahtaheri, S.J., Abdollahi, M., Golbabaei, F., Rahimi-Froushani, A. and Ghamari, F. (2004) Optimization of SPE for Analysis of Mandelic Acid as a Biomarker of Exposure to Ethyl Benzene. Iranian Journal of Environmental Health Science & Engineering, 1, 70-80.

13. Shahtaheri, S.J., Khadem, M., Golbabaei, F. and Rahimi-Froushan, A. (2006) Pre-Concentration of Cadmium Using Amberlite XAD-4 Prior to Atomic Absorption Spectroscopy. Iranian Journal of Environmental Health Science & En- gineering, 3, 45-52.

14. Shahtaheri, S.J., Khadem, M., Golbabaei, F. and Rahimi-Froushani, A. (2007) Optimization of Sample Preparation Procedure for Evaluation of Occupational and Environmental Exposure to Nickel. Iranian Journal of Public Health, 36, 73-81.

15. Shahtaheri, S.J., Ghamari, F., Golbabaei, F., Rahimi-Froushani, A. and Abdollahi, M. (2005) Sample Preparation Fol- lowed by High Performance Liquid Chromatographic (HPLC) Analysis for Monitoring of Muconic Acid as a Bio- marker of Occupational Exposure to Benzene. International Journal of Occupational Safety and Ergonomics, 11, 337-388.

16. Shahtaheri, S.J., Ibrahimi, L., Golbabaei, F. and Hosseini, M. (2007) Solid Phase Extraction for 1-Hydroxypyrene as a Biomarker of Exposure to PAHs Prior to High Performance Liquid Chromatography. Iranian Journal of Chemistry and Chemical Engineering, 26, 75-81.

17. Shahtaheri, S.J., Abdollahi, M., Golbabaei, F. and Rahimi-Froushani, A. (2008) Monitoring of Mandelic Acid as a Biomarker of Environmental and Occupational Exposures to Styrene. International Journal of Environmental Re- search, 2, 169-176.

18. Barnabas, I.J. and Dean, J.R. (1994) Owen Supercritical Fluid Extraction of Analytes from Environmental Samples, A Review. Analyst, 119, 2381-2394.

19. Viana, E., Molto, J.C. and Font, G. (1996) Optimization of a Matrix Solid-Phase Dispersion Method for the Analysis of Pesticide Residues in Vegetables. Journal of Chromatography A, 754, 437-444.

20. Barrionuevo, W.R. and Lancas, F.M. (2002) Comparison of Liquid-Liquid Extraction (LLE), Solid-Phase Extraction (SPE), and Solid-Phase Microextraction (SPME) for Pyrethroid Pesticides Analysis from Enriched River Water. Bulle- tin of Environmental Contamination and Toxicology, 69, 123-128.

21. Zhang, Z. and Pawliszyn, J. (1993) Headspace Solid Phase Microextraction. Analytical Chemistry, 65, 1843-1852.

22. Pan, L. and Pawliszyn, J. (1997) Derivatization/Solid-Phase Microextraction: New Approach to Polar Analytes. Analy- tical Chemistry, 69, 196-205.

23. Tsoukali, H., Raikos, N., Theodoridis, G. and Psaroulis, D. (2004) Headspace Solid Phase Microextraction for the Gas Chromatographic Analysis of Methyl-Parathion in Post-Mortem Human Samples. Forensic Science International, 143, 127-132.

24. Heidari, H.R., Shahtaheri, S.J., Golbabaei, F., Alimohamadi, M. and Rahimi-Froushani, A. (2008) Optimization of Headspace Solid Phase Microextraction Procedure for Trace Analysis of Toluene. International Journal of Occupa- tional Safety and Ergonomics (JOSE), 14, 395-405.

25. Heidari, H.R., Shahtaheri, S.J., Golbabaei, F., Alimohamadi, M. and Rahimi-Froushani, A. (2009) Trace Analysis of Xylene in Occupational Exposures Monitoring. Iranian Journal of Public Health, 38, 89-99.

26. Hernandez, F., Pitarch, E., Beltran, J. and Lopez, F.J. (2002) Headspace Solid-Phase Microextraction in Combination with Gas Chromatography and Tandem Mass Spectrometry for the Determination of

Organochlorine and Organophos- phorus Pesticides in Whole Human Blood. Journal of Chromatography B, 769, 65-77.

27. Kin, C.M. and Huat, T. (2010) Headspace Solid-Phase Microextraction for the Evaluation of Pesticide Residue Con- tents in Cucumber and Strawberry after Washing Treatment. Food Chemistry, 123, 760-764.

28. Alvarez, M.F., Llompart, M., Lamasa, J.P., Lores, M., Garcia-Jares, C., Cela, R. and Dagnac, T. (2008) Development of a Solid-Phase Microextraction Gas Chromatography with Microelectron-Capture Detection Method for a Multiresi- due Analysis of Pesticides in Bovine Milk. Analytica Chimica Acta, 617, 37-50.

29. Beceiro-Gonzalez, E., Concha-Grana, E., Guimaraes, A., Goncalves, C., Muniategui-Lorenzo, S. and Alpendurada, M.F. (2007) Optimisation and Validation of a Solid-Phase Microextraction Method for Simultaneous Determination of Different Types of Pesticides in Water by Gas Chromatography-Mass Spectrometry. Journal of Chromatography A, 1141, 165-173.

30. Namera, A., Watanabe, T., Yashiki, M., Iwasaki, Y. and Kojima, T. (1999) Simple Analysis of Arylamide Herbicides in Serum Using Headspace-Solid Phase Microextraction and GC/MS. Forensic Science International, 103, 217-226.

31. Tsoukali, H., Theodoridis, G., Raikos, N. and Grigoratou, I. (2005) Solid Phase Microextraction Gas Chromatographic Analysis of Organophosphorus Pesticides in Biological Samples. Journal of Chromatography B, 822, 194-200.

32. Yao, Z.W., Jiang, G.B., Liu, J.M. and Cheng, W. (2001) Application of Solid-Phase Microextraction for the Determi- nation of Organophosphorous Pesticides in Aqueous Samples by Gas Chromatography with Flame Photometric Detec- tor. Talanta, 55, 807-814.

33. Bicchi, C., Cordero, C., Liberto, E., Sgorbini, B. and Rubiolo, P. (2007) Reliability of Fibres in Solid Phase Microex- traction for Routine

Analysis of the Headspace of Aromatic and Medicinal Plants. Journal of Chromatography A, 1152, 138-149.

34. Sousa, E.T., Rodrigues, F.M., Martins, C.C., de Oliveira, F.S., Pereira, P.A.P. and de Andrade, J.B. (2006) Multivariate Optimization and HS-SPME/GC-MS Analysis of VOCs in Red, Yellow and Purple Varieties of Capsicum chinese sp. Peppers. Microchemical Journal, 82, 142-149.

35. Sakamoto, M. and Tsutsumi, T. (2004) Applicability of Headspace Solid-Phase Microextraction to the Determination of Multi-Class Pesticides in Waters. Journal of Chromatography A, 1028, 63-74.

36. Doong, R.A., Chang, S.M. and Sun, Y.C. (2000) Solid-Phase Microextraction for Determining the Distribution of Six- teen US Environmental Protection Agency Polycyclic Aromatic Hydrocarbons in Water Samples. Journal of Chroma- tography A, 879, 177-188.

37. Fernandez Alvarez, M., Liompart, M., Pablo Lamas, J., Lores, M., Garcia-Jares, C., Cela, R. and Dagnac, T. (2008) Development of a Solid Phase Microextraction Gas Chromatography with Microelectron-Capture Detection Method for a Multiresidue Analysis of Pesticides in Bovine Milk. Analytica Chimica Acta, 617, 37-50.

38. Kataoka, H., Lord, H.L. and Pawliszyn, J. (2000) Applications of Solid Phase Microextraction in Food Analysis. Journal of Chromatography A, 880, 35-62.

39. Gorecki, T. and Pawliszyn, J. (1997) Effect of Sample Volume on Quantitative Analysis by Solid-Phase Microextrac- tion. Analyst, 122, 1079-1086.

40. Xua, X.Q., Yang, H.H., Wang, L., Han, B., Wang, X.R. and Lee, F. (2007) Analysis of Chloroacetanilide Herbicides in Water Samples by Solid-Phase Microextraction Coupled with Gas Chromatography-Mass Spectrometry. Analytica Chimica Acta, 591, 87-96.

CHAPTER 10

Ximenia Americana: Chemical and Spectral Studies of Extracts of Seeds. Analysis of Trimethylsilyl Derivatives by Gas Chromatography and Mass Spectrometry

Romézio Alves Carvalho da Silva*, Telma Leda Gomes de Lemos, Daniele Alves Ferreira, Francisco José Queiroz Monte

Programa de Pós-Graduação em Química, Department of Organic Chemistry e Inorganic, Federal University of Ceará, Fortaleza, Brazil

ABSTRACT

Among the numerous botanical species found in the Brazilian plant biome, Ximenia americana is widespread in the Northeast of Brazil, especially in the state of Ceará, where it is known as bush plum. Its bark, leaves and roots are used in popular medicine for the treatment of skin infections, hemorrhoids, stomach ulcers, gastric pain, among others. In this work, hexane and ethanolic extracts of X. americana seeds were subjected to silylation reactions followed by analysis of the silylated derivatives by gas chromatography-mass spectrometry (GC/MS). In the hexane extract, 18 substances were identified, where the most abundant and most important components were the Octadec-9-enoic (38.14%), (9Z,12Z)-Octadec-9,12-

dienoic (18.83%), Ethanedioic (8.21%) and (Z,Z,Z)-9,12,15-Octadecatrienoic (7.22%) acids. From the ethanol extract 18 substances were also identified having D-Sucrose (29.36%), L-Sorbose (9.19%), syllo-Inositol (8.34%) and D-Glucose (7.45%) as the main components. In addition, the presence of a steroid in hexane extract and triterpenes in ethanol extract was recorded. The constituents were identified using chromatographic and spectrometric methods, especially gas chromatography-mass spectrometry. To our knowledge, this is the first study of this nature from the oil of the X. american seeds.

KEYWORDS

Ximenia americana, Seeds, TMS Carbohydrates derivatives, GC/MS

1. INTRODUCTION

Brazil is considered the country with the greatest biological diversity. It houses about 14% of the world's plant diversity distributed in different biomes, mainly the Amazon region, the Atlantic Forest and Caatinga [1] . The "caatinga" is an exclusively Brazilian biome. It reaches an area of 850 square kilometers, occupying about 10% of Brazilian territory and is present mainly in the interior of Northeast Brazil and northern Minas Gerais state [2] . Basically is formed by xerófilas plants or adapted to dry conditions (found in semiarid northeast). It has been recorded about a thousand plant species in the "caatinga". According to experts, there may be double that amount. Importantly, the investigation in natural products assumed a great sense of urgency in response to the expanding human population and its consequent demand for food and good health. However, while many countries increase their investments in chemistry and pharmacology of natural products derived from plants, investments in Brazil are still quite insufficient in this area. The Ximenia genus (Olacaceae) comprises about 8 species [3] , Ximenia americana being the most common. It is popularly known as "ameixa do mato", "ameixa de espinho" and "ameixa da Bahia" and is widely distributed in northeast Brazil. A tea obtained from its barks has been used in popular medicine as cicatizing, adstringent and as an agent against excessive menstruation. As a powder, it treats stomach ulcers and the seeds are purgative [4] [5] . In previous phytochemical study [6] , the stem and roots of this plant were found to contain saponins, glicosydes, flavonoids, tannins, phenolics, alkaloids, quinones and terpenoids types. In addition, the plant is potentially rich in fatty acids and glycerides and the seeds contain derivatives cyanide. Concerning the fatty acids, in addition to common C_{16}, C_{18} and C_{22}, a distinctive feature is the presence of acetylenic fatty acids. Our continuing interest in the chemistry of X. americana led us to investigate its seeds, still little studied from the chemical and pharmacological aspects. The hexanic and ethanolic extracts of X. americana were found to contain besides triglycerides, various fatty acids, furanoses/ pyranoses carbohydrates, steroids and triterpenoids. The identification of components was made through chemical and spectral studies.

2. EXPERIMENTAL PART

2.1. General

GC/MS analysis were performed on Shimadzu GCMS model QP2010SE Plus using a (5%-phenyl)-dimethyl- polysiloxane Rtx®-5MS capillary column (30 m × 0.25 mm × 0.25 μm) with film thickness 0.1 μm; the temperatures of the injector and detector were 260°C and 300°C, respectively; column conditions: 60°C (0.5 min) to 260°C (5 min) at 6°C/min, then 12°C/min to 300°C (10 min) using He as carrier gas, flow rate 1.7 mL/min with split mode. The analysis with the mass detector was performed in the scan mode with analysis time of 52.21 minutes; EIMS data were recorded with electron impact ionization at 70 eV (1.5 KV voltage, analyzer quadrupole and ion source 200°C), in the range 47 - 600 Da. ^{1}H- and ^{13}C-NMR spectra were recorded on Brucker DRX 500 spectrometer in C_5D_5N and are reported in ppm relative to using tetramethylsilane (TMS) as internal standard; IR spectra were recorded on Perkin-Elmer model Spectrum 100 FT-IR using a Universal Attenuated Total Reflectance (UATR). All solvents and reagents were of analytical grade and obtained from commercial sources (Synth and Merck). Silica gel 70 - 230 mesh (Merck) was used for column chromatography; silica gel Merck Kieselgel 60 (2 - 25 mm silica gel) with fluorescent indicator on aluminum support paper (Merck F254) was used for thin-layer chromatography (TLC) and UV light (254 and 365 nm) and vanillin/ perchloric acid/EtOH solution under heating, as developing agents. Mp's were determined on digital Mettler Toledo, model FP62 apparatus and are uncorrected.

2.2. Plant Material

Seeds of X. americana were collected in Marrecas site, near the city of Acarape, Ceará, Brazil in 2014. A voucher specimen (N° 040411) is deposited at the Herbarium Prisco Bezerra of the Department of Biology, Federal University of Ceará, Fortaleza, Brazil.

2.3. Isolation of triterpenoids

On concentration in vacuo, the crude EtOH extract afforded a greenish-white precipitate, which was filtered. The precipitate after successive washes with MeOH resulted in a white solid (60 mg) amorphous soluble in pyridine, which was characterized as a triterpenes mixture: mp 209°C - 212°C.

2.4. Isolation of triglycerides

The hexane fraction (2.1 g) of hexane extract was subjected to column chromatography on silica gel, eluted with increasing concentrations of EtOAc in hexane. After thin layer chromatography, the fractions 17 - 20 eluted with hexane-EtOAc 4:3, was rechromatographed on silica gel column eluted with increasing concentrations of CH_2Cl_2 in hexane. Thin layer chromatography brought together all fractions 16 - 44 (eluted with hexane-di- chloromethane 1:1), which were then characterized as triglycerides.

2.5. Synthesis of trimethylsilyl Derivatives

The extracts were submitted to a silylation reaction according to a known methodology [7] with slight modifications. Hexane extract (2 mg) was treated with BSTFA (100 μL) and TMCS (1%) and the mixture was stirred for 1 h at 60°C, followed by analysis by GC/MS.

3. RESULTS AND DISCUSSION

In separate experiments, hexane and ethanol crude extracts of X. americana seeds were successively partitioned into hexane, dichloromethane, ethyl acetate and methanol. Considering one of the main objectives of this study, namely the identification of oil components of the seed X. americana, using mainly gas chromatography coupled to mass spectrometry (GC/MS) and, without knowing the types of compounds present (hydrocarbons, alcohols, organic acids, esters of fatty acids, steroids, sugars, among others), preparation of volatile derivatives has become a basic requirement. Thus, the crude hexane was subjected to silylation reaction with BSTFA/ TMCS followed by obtaining a chromatogram (GC) of the crude reaction product. The structures of the components corresponding to 18 peaks in the chromatogram of total ion were identified by GC/MS from the fragmentation pattern observed in their mass spectra, and by the comparison to the mass spectra reported in the literature.

Among the identified compounds (Table 1), most are fatty acids, especially the hexadecanoic (palmitic acid), (Z,Z,Z)-9,12,15-octadecatrienoic (α-linolenic acid), (9Z,12Z)-octadec-9,12-dienoic (linoleic acid) and octadec-9-enoic (oleic acid) acids, the latter as the major component. It is also worth reporting the presence of the steroid β-Sitosterol (3.4%).

The MS of the TMS derivatives of these acids exhibit a characteristic profile fragmentation. Thus, their MS in addition to peaks due to $[M]^+$ show peaks at m/z 73, 117, 132; 145 and 313 corresponding to fragments $[(CH_3)_3Si]^+$, $[(CH_3)_3SiCO_2]^+$, $[(CH_3)_3SiCO_2CH_3]^+$, $[(CH_3)_3SiCO_2C_2H_4]^+$and $[(CH_3)_2SiCO_2C_{15}H_{31}]^+$, respectively. In general, at 70 eV eims of these trimethylsilyl derivatives, m/z 73 $[TMS]^+$ is registered as the base peak. Taken as an example, the mass spectrum (Figure 1) of the component with retention time 24.0 (Table 1) which was identified as the palmitic acid (RCO_2TMS, $R = C_{15}H_{31}$) showed a molecular peak at m/z 328.

The loss of methyl radical to give the ion at m/z 313 (M-15), occurs mainly from the silyl group [8] because at this point, the cation is more stable (Scheme 1). The ions at m/z 145 and 132 were explained by the fission of the bond β,γ (relative to the carbonyl group) of the chain R with possible transfer of hydrogen via a mechanism involving four centers, and through the expected McLafferty rearrangement [9] [10] , respectively.

Also, the mass spectra of silylated fatty acids containing a double bond showed virtually the same behavior. Even those containing more than one double bond, but showing differences, exhibited a similar profile. However, the peaks corresponding to the fragments at m/z 117, 132 and 145 in these latter cases were much less intense.

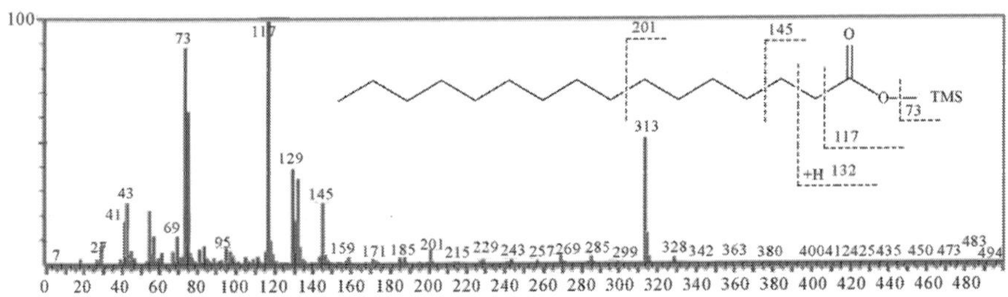

Figure 1. Mass spectrum of the TMS-hexadecanoic acid (palmitic acid).

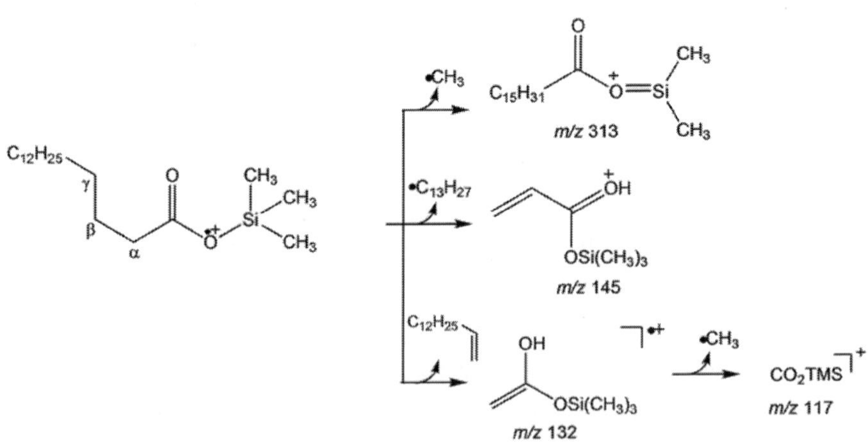

Scheme 1. Mass spectral fragmentation observed for TMS-hexadecanoic acid (palmitic acid).

Table 1. Components of the hexane extract of the seeds of *X. americana*.

Peak	Compounds	RT	%	Mass spectra, *m/z* (rel. intensity)
1	Ethanedioic acid	6186	8.21	73(29), 75(39), 90(28), 147(100)
2	Disiloxane, hexamethyl-	7100	5.10	66(12), 73(22), 147(100), 148(17)
3	Pentasiloxane, dodecamethyl-	8732	0.69	73(81), 147(100), 281(91), 369(18)
4	Glycerol	10,957	0.76	73(100), 103(18), 117(19), 147(49), 205(22), 218(14)
5	n.i.	12,230	1.89	73(66), 146(100), 205(62), 279(51), 367(100)
6	Nonadioic acid	20,403	0.97	55(46), 73(100), 75(76), 117(28), 129(30), 317(21)
7	Hexadecanoic acid	24,023	4.17	73(89), 75(62), 117(100), 129(40), 132(35), 145(26), 313(52), 328(5)
8	9,12-Octadecadienoic acid (*Z,Z*)-	26,297	18.83	73(100), 75(98), 117(20), 129(27), 132(4), 145(8), 337(39), 352(4)
9	cis-9-Octadecenoic acid	26,358	38.14	73(100), 75(85), 117(65), 129(70), 132(19), 145(33), 339(47), 354(6)
10	trans-9-Octadecenoic acid	26,440	1.70	73(100), 75(80), 117(80), 129(49), 132(15), 145(40), 339(67), 354(7)
11	Octadecanoic acid	26,666	2.17	73(79), 75(58), 117(100), 129(38), 132(37), 145(27), 341(50), 356(4)
11	Octadecanoic acid	26,666	2.17	73(79), 75(58), 117(100), 129(38), 132(37), 145(27), 341(50), 356(4)
12	Linolenic acid	27,887	7.22	73(80), 75(83), 79(100), 117(22), 129(19), 132(4), 145(10), 335(7), 350(3)
13	9,12-Octadecadiynoic acid	28,907	2.52	73(95), 75(100), 91(82), 105(84), 117(50), 119(58), 129(29), 132(7), 145(12), 148(57), 333(8), 348(4)
14	trans-Octadec-11-en-9-ynoic acid, 8-trimethylsilyloxy-, methyl ester	29,526	1.24	73(100), 147(13), 217(12), 237(90), 365(4)
15	1-Monooleoylglycerol	32,741	0.90	55(55), 73(94), 103(48), 117(18), 129(100), 147(71), 201(20), 203(28), 205(12), 265(9), 397(40), 410(5), 485(6), 500(3)
16	Tetracosanoic acid	33,449	0.58	73(100), 75(75), 117(79), 129(30), 132(34), 145(33), 201(12), 425(40), 440(8)
17	cis-15-Tetracosenoic acid	37,198	1.13	73(100), 75(82), 117(62), 129(68), 132(22), 145(34), 423(32), 438(8)
18	*β*-Sitosterol	39,343	3.04	43(58), 73(52), 129(100), 255(14), 357(42), 381(22), 396(40), 471(9), 486(14)
19	Lupeol	43,736	0.74	73(50), 189(100), 203(52), 218(38), 279(10), 369(12), 393(9), 408(5), 483(4), 498(9)

n.i.: no identified.

Steroids in general, exhibit a very similar fragmentation pattern. The mass spectrum of the constituent with retention time 39.343 in the chromatogram showed a molecular peak at m/z 486 and was identified as β-Sitosterol. The peaks at m/z 471, 396, 381, 357, 213 and 129, are the most representative, since they correspond to fragments of higher mass and are relatively more prominent peaks. The ion at m/z 471 (M-15) can occur in two paths (Scheme 2): fission of the C-Si bond (a) or the C_{10}-C_{19} bond (b). Studies [11] from similar compounds demonstrated that the most attractive site of methyl cleavage is at the C?Si bond (a), a type of fragmentation very common in the mass spectral fragmentation of TMS ethers and esters. The ion at m/z 396 (M-90) was justified due to the loss of trimethylsilanol (TMS-OH) with the formation of ionized diene (c). Studies with 2,2,4,4-d_4 steroids suggests that some of the migrant hydrogen comes from C-4 and on the basis of stereoche- mical considerations it seems most likely the 4β hydrogen is probably migrant [11] (Scheme 2). In addition, another fragmentation (d) related to the M?90 results in one fragment at m/z 75, for which high-resolution mass measurements [11] indicated the composition $(CH_3)_2SiOH$. This fragment results from cleavage of a methyl radical from the TMS function followed by loss of the dimethylsilanol moiety. Deuterium labeling and methyl substitution at C-4 reveal a situation identical with the hydrogen transfer involved in the M-90 fragmentation, namely, that only some (40%) of the hydrogen originates from the C-4 position and that it is probably the axial 4β hydrogen atom that is implicated [11] . In turn, the fragment at m/z 396 loses methyl radical to yield the m/z 381. According to the literature [10] , in the formation of m/z 357, the hydrogen atom H-2 it is transferred to C-6 carbon through six-atom system, followed by cleavage of the allylic bond C-3-C-4 and rearrangement of the ring A. Other typical ions (m/z 303, 275, 213, 184, 121, 95, 57, 43), resulting from the cleavage of the steroidal skeleton and of the side chain were observed [9] [11] [12] . The base peak at m/z 129 was justified by a complex fragmentation occurred in ring A. Labeling studies [11] -[13] with deuterium showed that the C_4 carbon is not part of ion m/z 129 (Scheme 2).

The crude EtOH extract was subjected to reaction silylation with BSTFA/TMCS followed by obtaining a chromatogram GC/MS of the crude reaction product. The structures of the components corresponding to 18 peaks in the chromatogram of total ion were identified by GC/MS from the fragmentation pattern observed in their mass spectra, and by the comparison to the mass spectra reported in the literature. Among the identified substances (Table 2), most are represented by furanosides and pyranosides carbohydrates and the main constituents were D-(+)-Sucrose (29.36%), L(-)-Sorbose (9.19%), scyllo-Inositol (8.34%) and D-Glucose (7.45%).

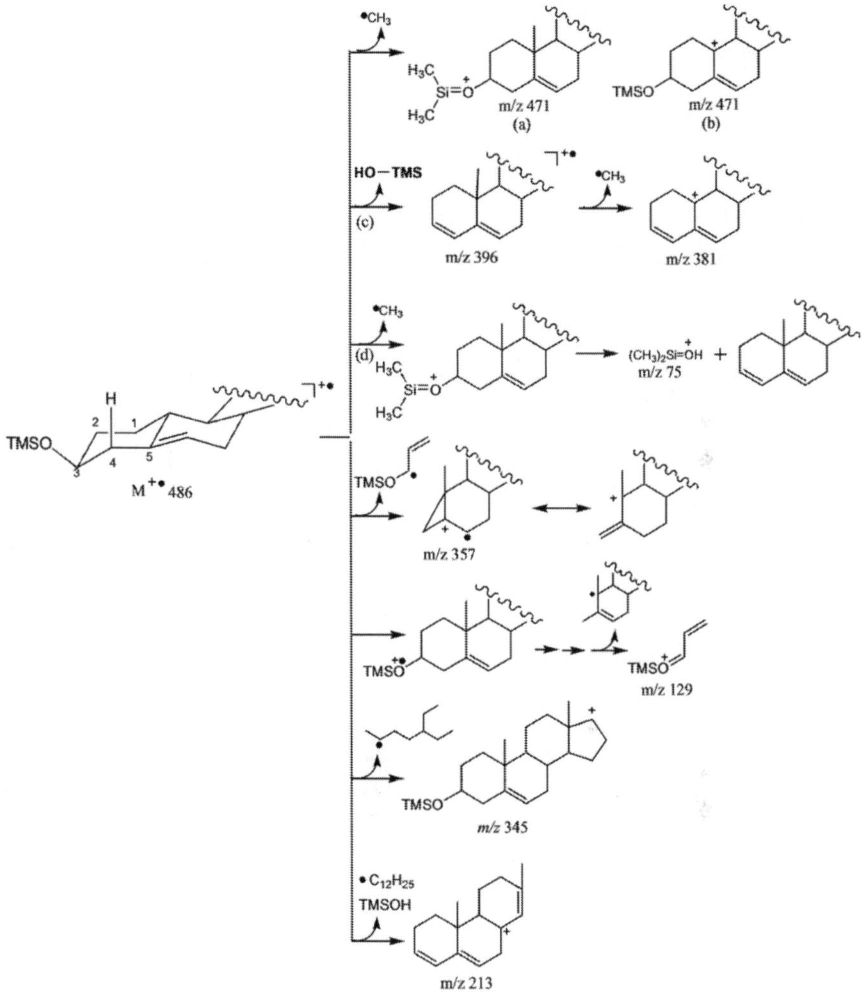

Scheme 2. Mass spectral fragmentation observed for TMS-β-Sitosterol.

The identification of these substances as trimethylsilyl derivatives by GC/MS can be made in view of some mechanistic proposals for characteristic fragments, according to published studies [14] -[16] . For example, the fragments m/z 204 and 217 are relatively common from mass spectra of trimethylsilyl derivatives of carbohydrates. Both peaks can be seen in mass spectra from TMS ethers, however, the peak at m/z 204 is generally the dominant from TMS ethers with pyranose ring, as observed in the spectrum (Figure 2) of the component with retention time 21.091 (Table 2). Its origin can occur by different mechanisms. One them, from the open-chain carbohydrate involves the TMS group via McLafferty type rearrangement [17] [18] the fragment being formed by the C-1?C-2 carbon (Scheme 3).

Table 2. Components of the EtOH extract of the seeds of X. americana.

Peak	Compounds	RT	%	Mass spectra, m/z (rel. intesity
1	1,2-Bis(trimethylsiloxy)ethane	6181	2.13	147(100), 73(58), 191(23)
2	Disiloxane, hexamethyl-	7097	1.46	66(12), 73(22), 147(100), 148(14)
3	Glycerol	10,958	5.58	73(100), 103(18), 117(19), 147(80), 205(38), 218(15)
4	n.i.	14,239	1.41	172(100), 82(94), 73(22)
5	beta.-D-Galactofuranoside	20,404	3.04	73(100), 217(82), 147(33)
6	D-Fructose	20,856	1.48	73(100), 217(55), 147(25)
7	Mannonic acid, lactone	20,982	3.08	73(100), 217(65), 147(27)
8	Sorbopyranose, L-	21,091	9.19	73(100), 204(68), 217(33)
9	D-Galactose	21,182	1.54	73(82), 204(100), 217(28), 147(51)
10	Inositol, scyllo-	21,305	8.34	305(20), 318 (32), 217(40), 147(42), 191(22), 204(12)
11	alpha.-D-Glucopyranose	22,212	5.15	204(100), 73(82), 191(40)
12	D-Mannitol	22,916	6.27	73(100), 147(51), 319(42)
13	D-Glucose	23,521	7.45	204(100), 191(41), 217(18)
14	Inositol, myo-	25,082	1.62	305(56), 318(30), 217(62), 147(51), 191(30), 204(25)
15	9,12-Octadecadienoic acid (Z,Z)-	26,297	1.91	73(94), 75(100), 129(25), 337(34)
16	trans-9-Octadecenoic acid	26,353	3.31	73(96), 75(100), 117(79), 129(68), 145(32), 339(42)
17	D(+)-Sucrose	31,944	29.36	361(100), 73(80), 217(32), 147(23),103(20), 437(12), 540
18	14,17-Nor-3,21-dioxo-.beta.-amyrin, 17,18-didehydro-3-dehydroxy-	43,022	1.79	73(100), 75(90), 117(55)
19	Silane, (9,19-cyclo-9.beta.-lanost-24-en-3. beta.-yloxy)trimethyl-	43,309	5.89	73(100), 117(62), 75(57)

n.i.: no identified

This two-carbon fragment, m/z 204, a common dominant peak of pyranose derivatives (aldo-, keto-, and glycosides), originates the most part, from C-2?C-3 and C-3?C-4 (aldopyranoses) [19] or must contain carbons C-3?C-4 and C-4?C-5 (ketopyranoses) [20] which may be formed from the molecular ion or from m/z 437, both by cyclic fragmentation [20] (Scheme 4). In turn, the ion at m/z 437, typical of ketoses [21] , results from the loss of C-1 and its substituent by fisson of the C-1?C-2 bond [19] [20] .

Thus, by comparing with published data [14] [15] , the mass spectrum of the component with RT 20.856 (Table 2) allow the assignment of a pyranose structure to the TMS ether derivative, which was identified as L-Sorbose in pyranose form. On the other hand, the relative intensity of the peak at m/z 204, which is most intense peak in the mass spectra of glucopyranosides (Figure 2) is drastically reduced in the mass spectra of glycofuranosides (Figure 3).

The peak at m/z 217 (Scheme 5) is characteristic of monosaccharides with furanose ring, which it is usually a very intense peak and, sometimes, the base peak. The relatively high intensity of this peak was observed in the mass spectrum of the component with retention time 20.856 (Table 2). Importantly, the fragment at m/z 217, according to the literature, is formed by carbon atoms C-2, C-3 and C-4 from aldopyranoses [19] [21] or from other carbons from ketofuranoses [18] , but without a specific proposal for fragmentation. On the other hand, its formation generally can occur from other parts of the molecule via rearrangement [19] . One possible mechanism involving rearrangement is seen in Scheme 5. In the spectrum (Figure 3) was not observed the molecular ion peak ($[M]^{+\bullet}$ 540), but was recorded peak at m/z 437 (M-CH_2OTMS), characteristic of 2-ketohexoses [20] [21] . Thus, by comparing the mass spectra obtained in this study with published data [14] [15], the component (RT 20.856) was identified as the TMS ether of D-Fructose in furanose form.

Finally, by comparing the mass spectrum (Figure 4) component at RT 31.944 with the mass spectra of the literature [16] , allowed to identify this component as the disaccharide Sucrose. It is noteworthy peaks at m/z 437 (M-CH_2OTMS), 361, 319, 217 and 204 (very low intensity) consistent with the pyranose and furanose structures. The other compounds of Table 2 were identified by comparison of retention times and mass spectral data reported in the literature.

An intense peak at m/z 306 which is generally observed in the mass spectra of carbohydrates in their open forms [21] [22] is not registered in the mass spectra analyzed in the present study. This observation is in agreement with the carbohydrates identified in their cyclic forms.

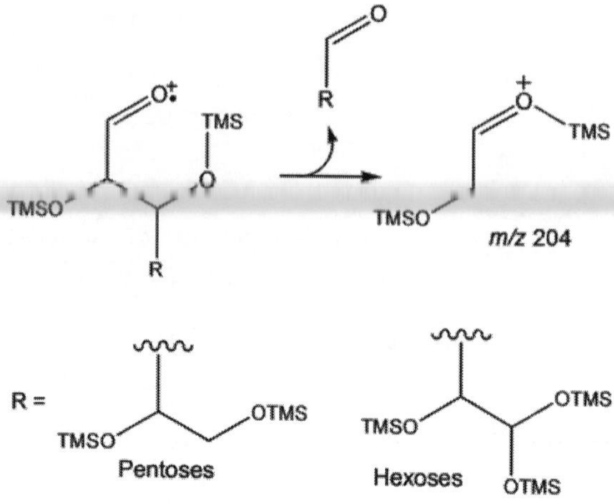

Scheme 3. Formation of fragment m/z 204 via McLafferty type rearrangement.

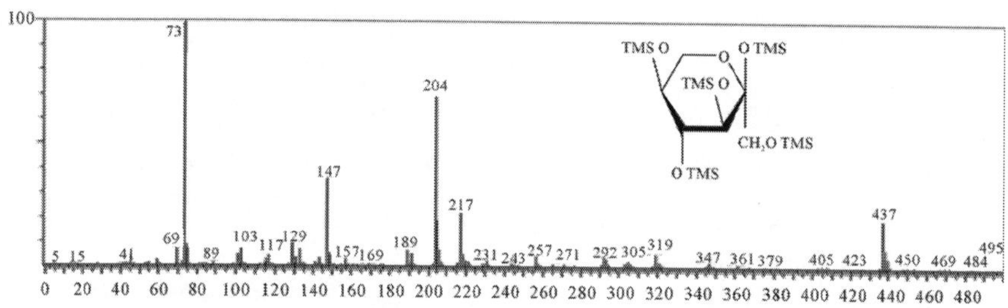

Figure 2. Mass spectrum of TMS-L-Sorbose.

At the concentration of the ethanolic extract of the seeds gave a precipitate which was recrystallized from MeOH. The ^{1}H NMR spectrum of this recrystallized (60 mg) showed that it was still mixture of two or three closely related compounds: IR spectrum γ max 3230 (OH), 2918/2910 (C_{sp3}-H), 1683 (C=O), 1029 (C?O) cm^{-1}. The ^{1}H NMR spectrum revealed the presence of various methyl groups, one of which (one singlet at 1.80 ppm) bonded to sp^2 carbon; the spectrum also showed signs as broad triplet at 5.72 (1H), broad singlet and 5.51 (1H) and two broad singlet at 4.96 (1H) and 4.78 (1H) due to hydrogens mutually coupled. The first two were attributed to olefinic hydrogens and consistent with the presence of triterpenes of oleanane/ursano types [23] , while the last two were correlated to vinyl hydrogens in terminal double bond, characteristic of triterpene lupane type [24] . In turn, the ^{13}C NMR spectrum in accordance with the ^{1}H NMR spectrum recorded signals to lupane [δ_C 151.68 (C-20) and 110.32 (C-29)], as well as signs consistent with oleanane triterpenes [δ_C122.92 (C-12), 56.10 (C-5), 42.36 (C-18), 46.83 (C-19), 33.54 (C-21) and 32.20 (C-22)] [23] -[25] ; the spectrum also showed signs at δ_C 78,44 (due to oxygenated sp^3 carbon C-3) and 19,12 correlated to methyl carbon C-30 in lupane triterpenoids [24] . In addition, the signals δ_C 180.56 and 179.23 characteristic of carbons of carboxylic acids with relative intensities of 1:2, respectively. The DEPT experiment confirmed the presence of hydrogenated carbon, especially oxygenated CH (δ_C 78.44), CH_2 sp^2 (δ_C 110.32), CH sp^2 (δ_C 122.92), CH sp^3 (δ_C 56.10 and 42.36) and CH_2 sp^3(46.83, 33.54 and 32.20). Based on the spectral comparison with the literature data (also considering the excessive number of signals in the ^{13}C NMR spectrum in the region between 14.00 and 56.00 ppm), it was concluded that the mixture was formed by triterpenes betulinic acid (56%), oleanolic acid (35%), and further a third component, also triterpenic nature in minor proportion (9%). The percentages were calculated from the integrations of signals at δ_H 3.56, 3.47 and 3.31 in the ^{1}H NMR spectrum, which were correlated with carbinolic hydrogens of the components of the mixture.

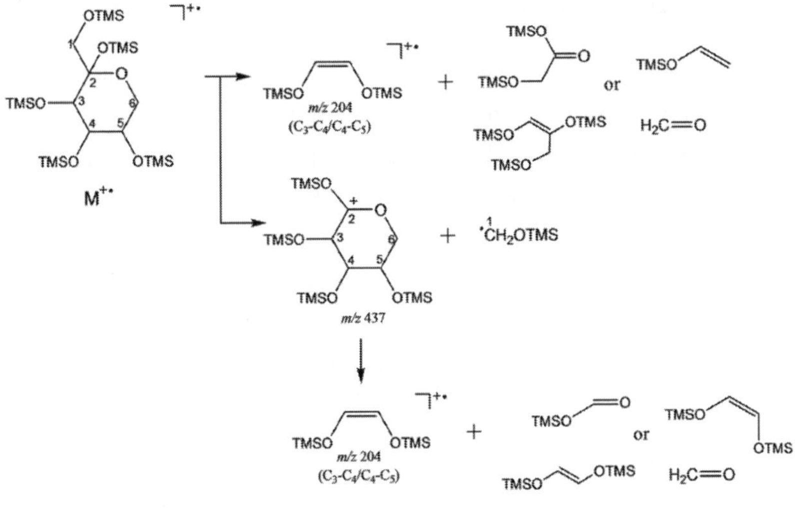

Scheme 4. Formation of fragment at m/z 204 via the alternative paths.

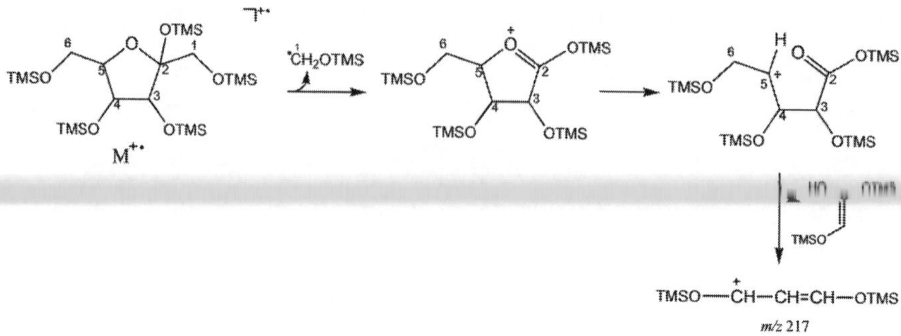

Scheme 5. Formation of fragment m/z 217 in furanose monosaccharides.

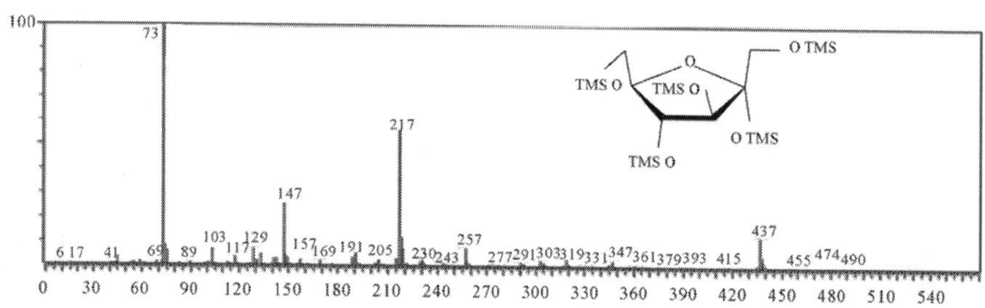

Figure 3. Mass spectrum of TMS-D-Frutose.

4. CONCLUSION

Carbohydrate analysis has proven to be a powerful technique in the understanding of biological processes. In addition to the possibility of studying the source of carbohydrates and their rotation in the environment, analysis of specific stable compounds may be used to trace microbial biomass and to detect food adulteration or in studies nutrition. The techniques used in the analysis have proved very useful, therefore, allowed the identification of various chemical constituents, mainly carbohydrates, from both crude extracts. To our knowledge, this is the first study from hexane and ethanol extracts of the seeds of X. ameriacana using silylated derivatives by gas chromatography-mass spectrometry. This study, the first with oil from the seeds of X. americana, enabled a possible use of this oil as human or animal nutraceutical ingredient.

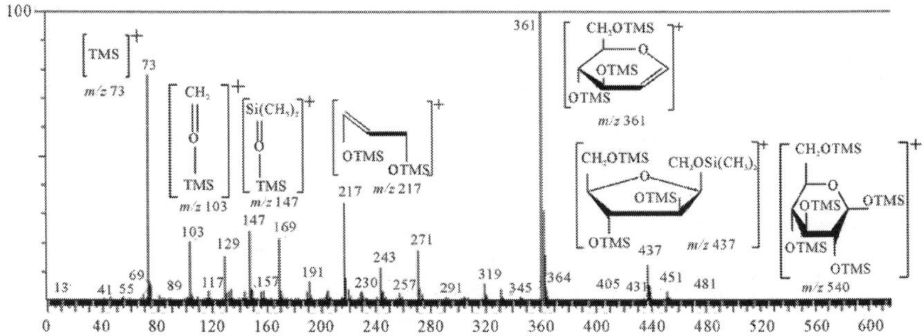

Figure 4. Mass spectrum of TMS-Sucrose.

ACKNOWLEDGEMENTS

The authors thank the Brazilian funding agencies FUNCAP, CAPES and CNPq for financial support.

REFERENCES

1. Sherpad, G.S. (2002) In: Lewinsohn, T.M. and Prado, P.I., Eds., Biodiversidade Brasileira: Síntese do Estado Atual de Conhecimentos, Contexto, St. Paulo, Brasil.

2. Alameida-Cortez, J.S., Cortez, P.H.M., Franco, J.M.V. and Uzunian, A. (2013) Caatinga. 2nd Edition, Colecao Biomas do Brasil, Ed. Harbra, St. Paulo.

3. Brasileiro, M.T., Egito, A.A., Lima, J.G., Randau, K.P., Perreira, G.C. and Neto P.J.R. (2008) Ximenia americana L.: Botanica, Química e Farmacologia no Interesse da Tecnologia Farmacêutica. Revista Brasileira de Farmácia, 89, 164-167.

4. Braga, R. (1976) Plantas do Nordeste, Especialmente do Ceará. 3nd Edition, Escola Superior de Agricultura, Mossoró.

5. Correia, M.P. (1984) Dicionário de Plantas úteis do Brasil e das Exóticas Cultivadas. Imprensa Nacional, Rio de Janeiro.

6. Monte, F.J.Q., Lemos, T.L.G., Araújo, M.R.S. and Gomes, E.S. (2012) Chapter 20. In: Rao, V., Ed., Phytochemicals, a Global Perspective of Their Role in Nutrition and Health, INTECH, Rijeka, Croatia, 429.

7. Isidorov, V.A. and Szczepaniak, L. (2009) Gas Chromatographic Retention Indices of Biologically and Environmentally Important Organic Compounds on Capillary Columns with Low-Polar Stationary Phases. Journal of Chromatography A, 1216, 8998-9007.

8. Clayden, J., Greeves, N., Warren, S. and Wathers, P. (2004) Organic Chemistry. Oxford University Press, Florence.

9. Budzikiewicz, H., Djerassi, C. and Williams, D.H. (1967) Mass Spectrometry of Organic Compounds. Holden-Day, London.

10. Silvério, F.O., Barbosa, L.C.A., Silvestre, A.J.D., Piló-Veloso, D. and Gomide, J.L. (2007) Comparative Study on the Chemical Composition of Lipophilic Fractions from three Wood Tissues of Eucalyptus Species by Gas Chromatography-Mass Spectrometry Analysis. Journal of Wood Science, 53, 533-540.

11. Diekman, J. and Djerassi, C. (1967) Mass Spectrometry in Structural and Stereochemical Problems. CXXV. Mass Spectrometry of Some Steroid Trimethylsilyl Ethers. Journal of Organic Chemistry, 32, 1005-1012.

12. Brooks, C.J.W. (1979) Some Aspects of Mass Spectrometry in Research on Steroids. Philosophical Transactions A, 293, 53-67.

13. Gustafsson, J.A., Ryhage, R., Sjoval, J. and Moriarty, R.M. (1969) Migrations of the Trimethylsilyl Group upon Electron Impact in Steroids. Journal of the American Chemical Society, 91, 1234-1236.

14. Chizhov, O.S., Molodtsov, N.V. and Kochetkov, N.K. (1967) Mass Spectrometry of Trimethylsilyl Ethers of Carbohydrates. Carbohydrate Research, 4, 273-276.

15. Medeiros, P.M. and Simoneit, B.R.T., (2007) Analysis of Sugars in Environmental Samples by Gas Chromatography-Mass Spectrometry. Journal of Chromatography A, 1141, 271-278.

16. Ferreira, C. (2011) Caracterizacao por GC-MS de Glícidos Derivatizacao Assistida por Microondas. Dissertation, Universidade de Coimbra, Coimbra.

17. Petersson, G. (1972) A McLafferty Type Rearrangement of a Trimethylsilyl Group in Silylated Hydroxy Carbonyl Compounds. Organic Mass Spectrometry, 6, 577-592.

18. Mogosanu, G.D., Grumezescu, A.M., Mihaiescu, D.E., Istrati, D., Mogosanu, D.E. and Buteica, S.A. (2011) Identification of Sugars from Silene albae herba Using GC-MS Technique. UPB Scientific Bulletin, Series B, 73, 101-108.

19. DeJongh, D.C., Radford, T., Hribar, J.D., Hanessian, S., Bieber, M., Dawson, G. and Sweeley, C.C. (1969) Analysis of Trimethylsilyl Derivatives of Carbohydrates by Gas Chromatography and Mass Spectrometry. Journal of the American Chemical Society, 91, 1728-1740.

20. Karady, S. and Pines, S.H. (1970) Mass Spectrometry of the Trimethylsilyl Ethers of 2-ketohexoses. Tetrahedron, 26, 4527-4536.

21. García-Raso, A., Fernández-Diaz, M., Páez, M.I., Sanz, J. and Castro, I.M. (1989) Gas Chromatographic Retention of Carbohydrate Trimethylsilyl Ethers. Ketohexoses. Journal of Chromatography, 471, 205-216.

22. Curtius, H.-Ch., Müller, M. and Vollmin, J.A. (1968) Studies on the Ring Structures of Ketoses by Means of Gas Chromatography and Mass Spectroscopy. Journal of Chromatography, 37, 216-224.

23. Henrique, M.C., Nunomura, R.C.S., Nunomura, S.M. and Silva, S.G. (2014) Constituintes Químicos Isolados dos Galhos e Cascas de Ampazeiro (Parahancornia amapa, Apocynaceae). Acta Amazonica, 44, 533-538.

24. Mahato, S.B. and Kundu, A.P. (1994) 13C NMR Spectra of Pentacyclic Triterpenoids—A Compilation and Some Salient Features. Phytochemistry, 37, 1517-1575.

25. Costa, D.A., Chaves, M.H., Silva, W.C.S. and Costa, C.L.S. (2010) Constituintes Químicos, Fenóis Totais e Atividade Antioxidante de Sterculia striata St. Hil. et Naudin. Acta Amazonica, 40, 207-212.

CHAPTER 11

Gas chromatographic–mass spectrometric determination of azithromycin in biological fluids

Subbiah Thangadurai

Post Graduate Studies and Research Department of Chemistry, Raja Doraisingam Government Arts College, Government of Tamil Nadu, 630 561 Sivagangai, India

ABSTRACT

Background

Azithromycin monohydrate is an antibiotic, which belongs to the category of macrolide antibiotics. It is used for an effective antibiotic for the treatment of sexually transmitted diseases, upper and lower respiratory tract infections and skin structure infections.

Methods

A gas chromatographic–mass spectrometric (GC–MS) method is described for the determination of azithromycin monohydrate residue in biological fluids. This method allows detection of residual azithromycin (AZM) in biological fluids by using single-ion monitoring (SIM). The confirmation by a full-scan electron impact (EI) mass spectrum is possible if residual level in a sample is >5 μg/mL. AZM is extracted with chloroform from a sample and cleaned up by n-hexane washing followed by partition between chloroform and phosphate buffer solution. The cleaned up extract is acetylated in acetic anhydride–pyridine mixture (1:2) at room temperature.

Results

The reaction mixture is injected into the GC–MS apparatus, and the detection has been conducted using SIM at m/z 200. The detection limit is 2 μg/mL.

Conclusions

This proposed method could be successfully applied to the drug analysis in the forensic laboratories as well as in the toxicological laboratories.

KEYWORDS

Azithromycin Biological fluids Extraction Derivatization GC–MS SIM

1. BACKGROUND

Antibiotics are now used extensively in food-producing animals to maintain optimal health and to promote growth. The use of these drugs can leave drug residues in edible tissues. These drug residues may have direct toxic effects moreover; safety considerations may include allergic reactions of consumers and induction of resistant bacteria. Therefore, conditions of use must be established for animal drugs to assure human food safety as well as efficacy and safety to the animal species (Livingston 1985).

Azithromycin (AZM), with a different chemical entity (so-called azalide) from the macrolides, is also derived from erythromycin, but the lactone ring is expanded from 14-membered to 15-membered, by the incorporation of nitrogen. As in erythromycin, cladinose and desosamine sugar residues are attached at positions 3 and 5. AZM exhibits a more extensive spectrum of activity, greater acid stability and more favourable pharmacokinetic parameters than erythromycin.

AZM is an effective antibiotic for the treatment of sexually transmitted diseases, upper and lower respiratory tract infections and skin structure infections (Kirst and Sides 1989). It is also used primarily to treat various bacterial infections caused by respiratory pathogens, such as aerobic Gram-positive and Gram-negative bacteria. It prevents bacterial cells from manufacturing specific proteins necessary for their survival. AZM is rapidly absorbed and is widely distributed to tissues and becomes concentrated in cells. Peak plasma concentrations are achieved (Debremaeker et al. 2003) within 2 to 3 h.

AZM is obtained from erythromycin, impurities present will undergo the same modifications and the AZM analogues of these impurities can be found in AZM bulk samples. In addition, degradation products of AZM as well as intermediate compounds (Khedr and Sheha 2003) of the semi-synthesis may be present; for instance, AZM is rapidly decomposed in acidic solution via intra-molecular dehydration to form erythromycin-6, 6-hemiketal and then anhydroerythromycin (Fiese and Steffen 1990). It is also very difficult to determine (Miguel and Barbas2003) small amounts of degradation products in a vast excess of parent drug and even more so when the compounds do not present a chromophore as this makes their detection more difficult.

Chromatographic methods using both thin-layer chromatography (TLC) and liquid chromatography (LC) have been described for the determination of macrolide antibiotics. The application of these methods to residue analysis has been limited. To overcome this lack of sensitivity and selectivity, TLC and LC methods have been developed; a review was found on these methods (Moats 1985). However, TLC methods are capable of qualitative detection of only one drug in one type of food. LC has also been used for determination of macrolide antibiotics, but application to residue analysis has been limited. A stability-indicating TLC method of the analysis of AZM in bulk and a capsule form is developed (Khedr and Sheha 2003).

Both coulometric and amperometric methods for the detection of AZM in human and animal tissues and serum were used (Shepard et al.1991). The coulometric method has since been used by a number of investigators to characterize the pharmacokinetics of AZM in humans (Ripa et al. 1996;

Conte et al. 1996). HPLC with ultraviolet detection has been used for the analysis of AZM in bulk samples, for the separation of related compounds produced during synthesis and for acid degradation studies. Methods for the HPLC analysis of AZM in biological samples have also been described using various methods of detection in order to overcome the limitations of poor UV absorbance.

A sensitive HPLC assay using a C_{18} stationary phase with detection by atmospheric pressure chemical ionization–mass spectrometry (APCI–MS) has been developed (Fouda and Schneider 1995), specifically for the analysis of AZM in children, which requires only a minute sample volume of 50 μL. A selective reversed-phase liquid chromatography/mass spectrometry (LC/MS) method has been developed for the identification of AZM impurities and related substances in commercial AZM samples (Debremaeker et al. 2003). In comparison with UV detection, this hyphenated technique provides as its main advantage efficient identification of novel substances without time-consuming isolation and purification procedures.

The purpose of this work is to briefly describe mass spectrometry techniques that are commonly used to solve toxicological problems. Often a gas chromatographic–mass spectrometric (GC–MS) quantitative assay is developed and used as a primary standard for evaluation of less expensive assay methods. Most GC–MS quantitative analyses involve the operation of the mass spectrometer in the selected ion monitoring mode (Falkner et al. 1975).

Methods used to determine AZM residue levels include, but are not limited to, microbiological and chromatographic techniques such as TLC and GC. However, microbiological methods lack specificity and TLC methods lack sensitivity. Microbiological assays are most commonly used for the measurement of AZM in food, but they are time consuming, they cannot identify certain macrolides and their precision appears to be variable. Therefore, a precise chromatographic analysis method for the macrolides has been required.

This paper describes a GC–MS method for the determination of AZM residue in biological fluids. This method allows detection of residual AZM in biological fluids by using single-ion monitoring. This method could be applied in the forensic laboratories and toxicological laboratories as well as in pharmaceutical formulations.

2. METHODS

2.1. Chemicals and materials

All reagents and solvents used were of analytical grade.

Azithromycin was received from Hindustan Antibiotics Ltd., Bangalore, India, as a free gift sample. Acetic anhydride and anhydrous pyridine were obtained from Merck (Darmstadt, Germany) and was pro analysis grade. De-ionized, doubly distilled water was used throughout.

TLC silica gel F_{254} plate was obtained from Merck (Darmstadt, Germany).

Standard reagents viz. methanol (CH_3OH), *n*-hexane (n-C_6H_6), chloroform ($CHCl_3$), sodium hydroxide (NaOH), sodium chloride (NaCl), sodium sulfate (Na_2SO_4) and disodium hydrogen phosphate ($Na_2HPO_4 \cdot 12H_2O$) all used were AR grade.

2.2. Na_2HPO_4 solution

One gram of disodium hydrogen phosphate ($Na_2HPO_4 \cdot 12H_2O$) was taken from a beaker and then dissolved in 100 mL of doubly distilled water.

2.3. NaOH solution

A 5 N sodium hydroxide (NaOH) solution was prepared using 20 g of NaOH and dissolved in 100 mL of doubly distilled water.

2.4. NaCl solution

Ten grams of sodium chloride (NaCl) was taken from a beaker and dissolved in 100 mL of doubly distilled water.

2.5. Standard solutions

2.5.1. Stock solution

Accurately weighed 100 mg of pure azithromycin monohydrate was dissolved in the methanol; the solution was made up in the 10 mL standard flask to make the concentration 10 mg/mL.

2.5.2. Working solution

Standard solutions were diluted to 50, 25, 10, 5 and 2 µg/mL concentration with methanol from the above stock solution. These solutions have been stored in a refrigerator with light protection.

2.6. Extraction and cleanup

Gastric cleavage samples (10 mL) were taken from a 100-mL centrifuge tube. Then, 20 mL of CH_3OH was added into that, and then it has been homogenized for 10 min. The contents were rinsed twice well with 3 mL of CH_3OH, and the rinses were added into the centrifuge tube.

The centrifuge tube has been shaken very well for 10 min. The supernatant solution was filtered through fluted paper into a 100-mL separatory funnel. Twenty milliliters of CH_3OH was added into the residue; the centrifuge tube was shaken vigorously for 10 min.

The contents were centrifuged for 10 min at 2000 rpm; after that, the supernatant solution was filtered through fluted paper. The filtrate was combined with the first extract, which is in a 100-mL separatory funnel.

Further, 20 mL of *n*-hexane was added into the combined CH_3OH extract, which is in a separatory funnel, it has been shaken vigorously for 5 min, and it was set aside to separate.

The lower CH_3OH phase was drained into a second 100-mL separatory funnel, and the upper *n*-hexane phase was discarded. Two to three milliliters of 1 N NaOH and 30 mL of $CHCl_3$ were added into the CH_3OH phase, and these were mixed by swirling.

In addition to this, 30 mL of 1 % of Na_2HPO_4 solution was also added, and it has been shaken vigorously for 2–3 min. It was set aside to separate for 15–20 min. The lower $CHCl_3$ phase was drained into a third 100-mL separatory funnel.

Suppose if two phases do not readily separate after centrifugation for 5 min at 2000 rpm, we have to add an additional 30 mL of $CHCl_3$ into the aqueous phase and it should be extracted again. The $CHCl_3$ extract was combined into the third 100-mL separatory funnel, and it was washed with 30 mL of NaCl solution.

The $CHCl_3$ phase has been transferred into a 100-mL round bottom flask, and the solvent has been evaporated at 40–45 °C.

2.7. Instrumentation

2.7.1. GC column

The capillary column was 5 % phenyl methyl silicone, with 30-m width, 0.25-mm i.d. and 0.25 film thickness.

2.7.2. Gas chromatograph mass spectrometer

The gas chromatograph mass spectrometer was a PerkinElmer instrument with Autosystem XL GC, turbo mass MS, autosampler, 2-μL injection, 200° injection port, 200° interface, 70-eV electron impact (EI) mode and 40–630-amu mass range.

2.8. Operating conditions

The operating conditions were as follows: 200° injection port, 250°C oven isothermal temperature, 1 mL/min carrier gas helium flow and 70-eV ionization voltage.

2.9. Derivatization procedure

For successful GC–MS analysis, the sample must be both volatile and thermally stable. Many organic compounds do not fulfil these criteria by virtue of their possessing polar functional groups. Protecting these functional groups by derivatization is the normal method for conferring good GC–MS behaviour. Derivatization is used to increase a compound's volatility or its thermal stability to improve its GC behaviour by substituting lipophilic groups for 'active' hydrogens or to advantageously alter the compound's mass spectrum (Lin et al. 2008).

The derivatization reactions generally used for GC determination are silylation, acetylation and etherification. In this work, we have chosen to utilize the acetyl derivative, because it is easily prepared and stable, and the increase in molecular weight by derivatization is not so large. Azithralosamine was acetylated using acetic anhydride and pyridine at room temperature. The reaction mixtures were monitored by TLC to ensure the complete conversion of starting materials. At high temperatures, it was possible to elute azithralosamine acetate as a relatively sharp peak in a reasonable retention time. In an autosampler vial, 1 mL of working solution containing a concentration of 5 μg/mL was pipetted out. It was evaporated to dry, and the dry residue was dissolved in 500 μL of anhydrous pyridine; in addition to the pyridine solution, 250 μL of acetic anhydride was added and the vial was closed tightly. The residue was dissolved by swirling and allowed to stand at room temperature overnight. This sample solution has been reserved in a refrigerator. This is ready for injection into the GC–MS apparatus.

2.10. Evaluation of repeatability, reproducibility and linearity

The repeatability of the derivatization procedure and the chromatographic analysis was determined by five replicate injections of the standards prepared as described above. The reproducibility of the method was determined by five replicate injections performed within 3 days.

The reproducibility of the derivatization procedure and the chromatographic analysis was checked by repetition of the first average separate determinations. The calibration curves were linear in the tested range up to 500 μg/mL.

3. RESULTS AND DISCUSSION

The molecular weight is relatively large (M.W=749) for AZM. This antibiotic has polar functional groups such as hydroxyls, carbonyls and sugars. Elution from a GC column after derivatization was considered to be difficult and impractical. At first, we intended to obtain the aglycone of AZM by hydrolyses of all glycoside linkages. However, the glycoside linkage of the amino sugar, desosamine, resists acid hydrolysis much more than that of the neutral sugar, cladinose. A condition normally required for the removal of desosamine from the macrolides induces destruction of the aglycone. On the other hand, azithralosamine can be easily obtained by mild acid hydrolysis (Fig. 1) encouraging us to do GC–MS determination after derivatization.

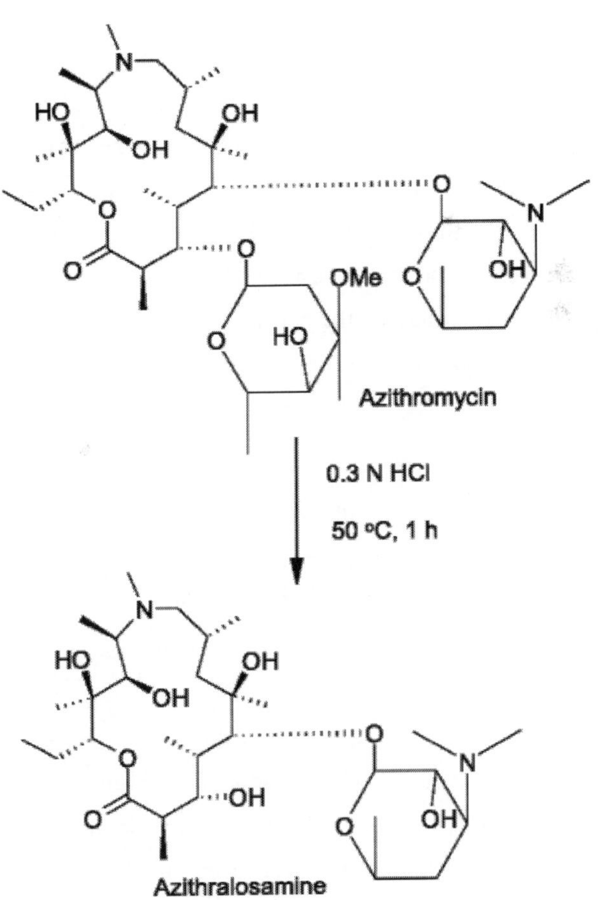

Figure 1. Acid hydrolysis of azithromycin

The conditions of acid hydrolysis for AZM was investigated using TLC (silica gel F_{254}, dichloromethane:methanol = 9:1). But no separation for AZM drugs was obtained. For this purpose (Wallhäuser KH 1967), a slow moving solvent mixture containing water and *n*-butanol had to be used, needing about 1 h to give an elution zone of 6 cm. With longer distances, separation of AZM was not improved. In contrast to dichloromethane:methanol mixtures, the R_f values of AZM spots did not change when coextractives from tissues were present. The R_f values with *n*-butanol:water (9:1 *v*/*v*) for AZM is 0.42. Optimum conditions obtained were 0.1 N HCl at room temperature and overnight reaction time. To shorten the reaction time, slightly more vigorous conditions were tried. We found that heating at 50 °C in 0.3 N HCl for 1 h was sufficient to obtain comparable yield of azithralosamine.

The reproducibility of the derivatization procedure and the chromatographic analysis was checked by repetition of the five average separate determinations. The calibration curves were linear in the tested range up to 500 µg/mL. The acetylation at an elevated temperature is faster than that at room temperature. The amount of acetylated product decreased to about half after 2 weeks of storage in the refrigerator. Our experiences indicated that the GC–MS determination must be performed within 2–3 days after acetylation.

Azithralosamine acetate was injected into the GC–MS apparatus, and total ion chromatogram and a full-scan mass spectrum were obtained. Figure 4 shows the mass spectrum, in which the base peak of *m*/*z* 200 is very prominent and *m*/*z* 84, 98, 123, 140, 158, 193, 207, 221, 223, 292, 401, 417, 487, 537 and 592 are observed but relatively weak. The *m*/*z* 200 ion is considered to be the desosamine acetate ion (Fig. 2). Simultaneous monitoring of three ions is usually regarded as a minimum for identification by selected ion monitoring of GC–MS spectra (Sphon 1978). In this work, we attempted to detect azithralosamine acetate using selected ion monitoring at *m*/*z* 200, 98, 123, 140, 158, 193, 207 and 221. However, the sensitivity of detection is much lower (<1/10) than that using single-ion monitoring (SIM) at *m/z* 200 only. There are no interfering peaks on this chromatogram at the retention time of azithralosamine acetate.

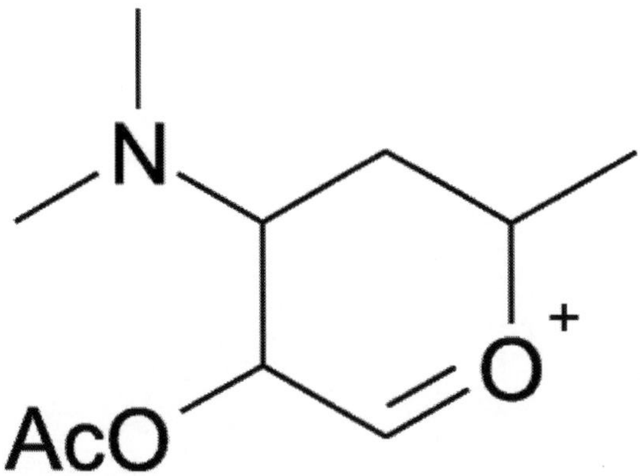

Figure 2. Structure for *m/z* 200 fragmentation of desosamine acetate in EI mass spectrum of the azithralosamine acetate

For extraction of AZM from sample tissues, we used methanol because of the high solubility of macrolides and the ease of permeability into sample tissues. To remove coextractives and to clean up the antibiotic, we washed methanol extract with *n*-hexane and partitioned between chloroform and phosphate buffer. Figure 3 shows the SIM chromatogram of AZM hydrolysis and acetylation at *m/z* 200. The peak observed is very close to that of azithralosamine acetate. Figure 4 shows the full-scan mass spectrum of the peak indicated at a retention time of 18.43. However, many background ions appeared on the full-scan mass spectrum of this peak and confirmation was not possible.

4. CONCLUSIONS

Although an internal standard is not used in this method, the separation of the GC column and the sensitivity of SIM detection considered together offer the reliable screening of residual AZM in biological fluids, and a confirmation of the presence of AZM in samples of high residue level is possible. The focus of our research however has been on methods allowing the confirmation of AZM in biological fluids. This method could be applied in the determination of macrolide group of antibiotics in forensic samples as well as in toxicological and in the environmental samples.

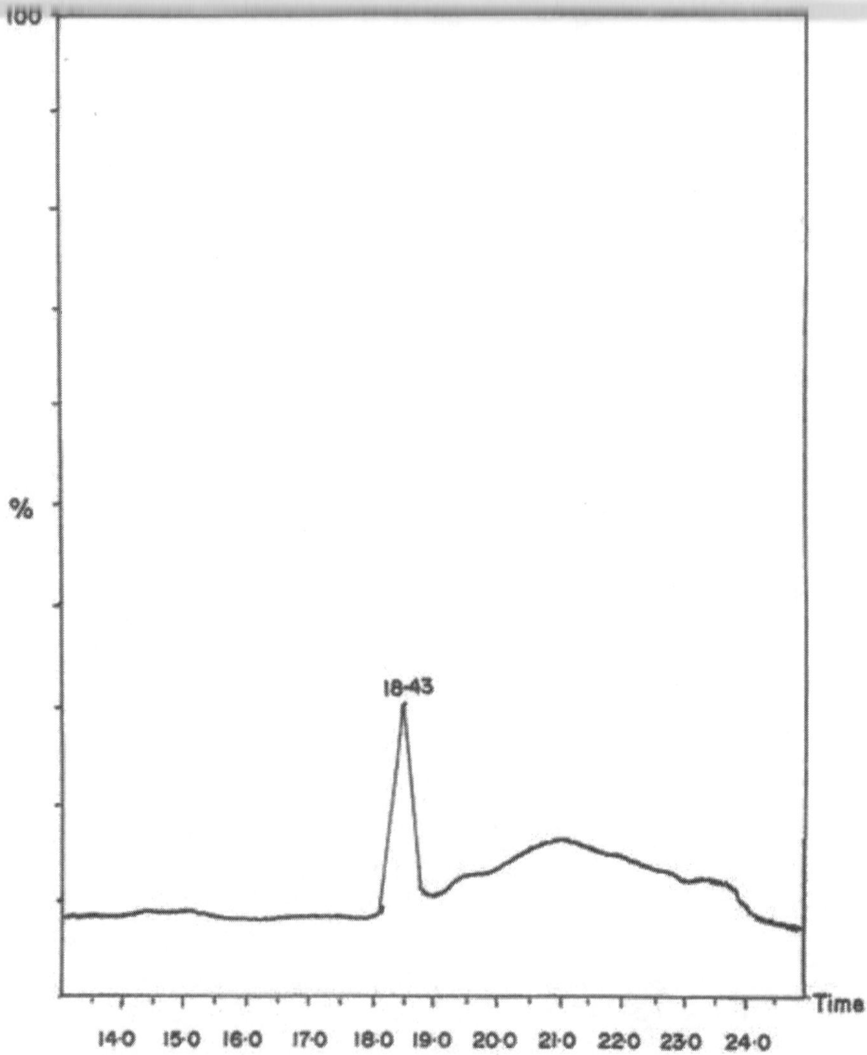

Figure 3. Single-ion chromatogram at *m*/*z* 200 for azithralosamine acetate

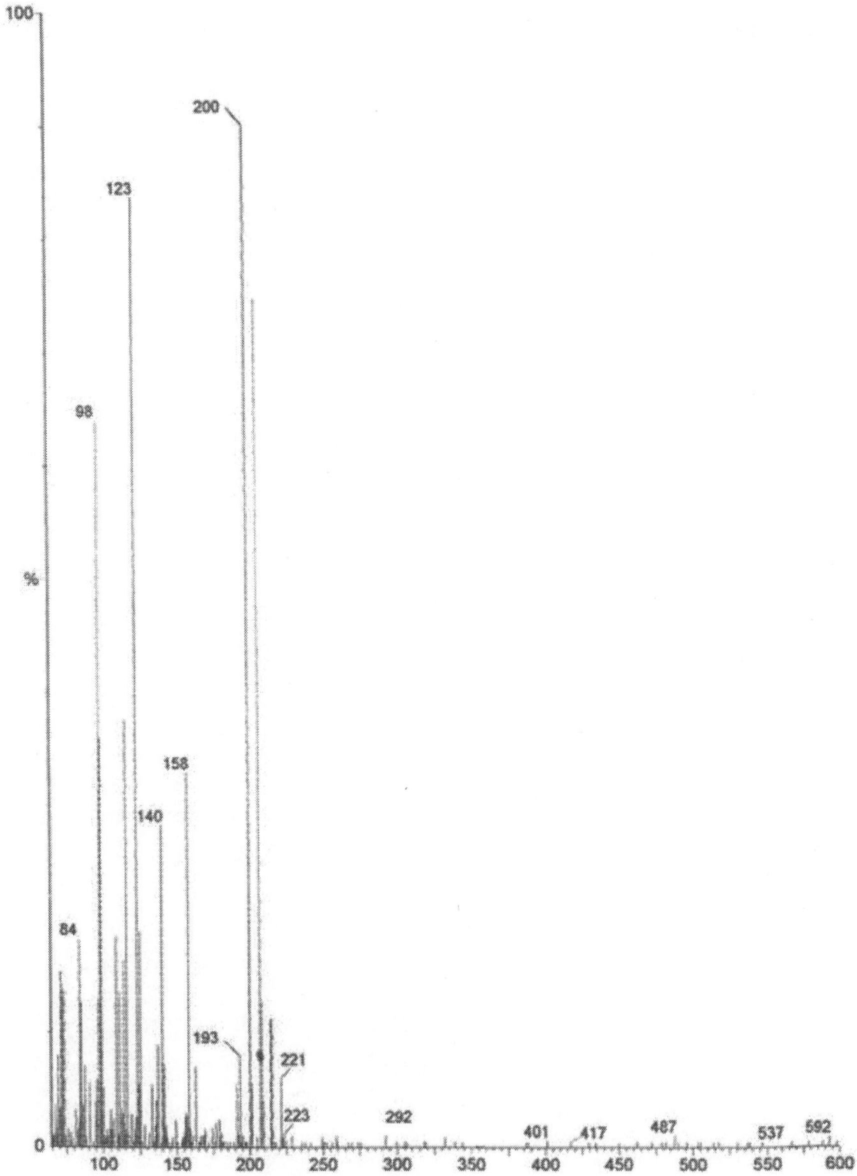

Figure 4. Electron impact mass spectrum of azithralosamine acetate (70 eV) at a retention time of 18.43 min

ACKNOWLEDGEMENTS

I am very grateful to the Director, Central Forensic Science Laboratory (CFSL), Directorate of Forensic Science (DFS), Ministry of Home Affairs, Govt. of India, CFI's Complex, Ramanthapur, Hyderabad 500 013, India, for providing the necessary facilities to carry out this work and for giving constant encouragement throughout the work.

AUTHORS' CONTRIBUTIONS

I have performed the experimental and analytical work and prepared the draft of the manuscript. The guidelines and supervision of this work was provided by the Central Forensic Science Laboratory, Directorate of Forensic Science (DFS), Ministry of Home Affairs, Govt. of India, CFI's Complex, Ramanthapur, Hyderabad 500 013, India. I have read and approved the final manuscript.

AUTHORS' INFORMATION

My research areas of interest are geochemistry, pharmaceutical and medicinal chemistry and forensic chemistry.

REFERENCES

1. Conte JE, Golden J Jr, Duncan S, McKenna E, Lin E, Zurlinden E (1996) Single-dose intrapulmonary pharmacokinetics of azithromycin, clarithromycin, ciprofloxacin, and cefuroxime in volunteer subjects. Antimicrob Agents Chemother 40(7):1617–1622

2. Debremaeker D, Visky D, Chepkwony HK, Van Schepdael A, Roets E (2003) Analysis of unknown compounds in azithromycin bulk samples with liquid chromatography coupled to ion trap mass spectrometry. Rapid Commun Mass Spectrom 17:342–350

3. Fouda HG, Schneider RP (1995) Quantitative determination of the antibiotic azithromycin in human serum by HPLC-atmospheric

pressure chemical ionization mass spectrometry: correlation with a standard HPLC electrochemical method. Ther Drug Monit 17:179–183

4. Falkner FC, Sweetman BJ, Watson JT (1975) Biomedical application of selected ion monitoring. Appl Spectrosc Rev 10:51–57
5. Fiese EF, Steffen SH (1990) Comparison of the acid stability of azithromycin and erythromycin A. J Antimicrob Chemother 25(Suppl A):39–47
6. Khedr A, Sheha M (2003) Quantitative thin layer chromatographic method of analysis of azithromycin in pure and capsule forms. J Chromatogr Sci 41:10–16
7. Kirst HA, Sides GD (1989) New directions for macrolide antibiotics: pharmacokinetics and clinical efficacy. Antimicrob Agents Chemother 33:1419–1422
8. Livingston RC (1985) Antibiotic residues in animal-derived food. J Assoc Off Anal Chem 68(5):966–967
9. Lin DL, Wang SM, Wu CH, Chen G, Liu RH (2008) Chemical derivatization for the analysis of drugs by GC-MS-A conceptual review. J Food Drug Anal 16(1):1–10
10. Miguel L, Barbas C (2003) LC determination of impurities in azithromycin tablets. J Pharm Biomed Anal 33:211–217
11. Moats WA (1985) Chromatographic methods for determination of macrolide antibiotic residues in tissues and milk of food-producing animals. J Assoc Off Anal Chem 68(5):980–984
12. Ripa S, Ferrante SL, Prenna M (1996) A linear model for the pharmacokinetics of azithromycin in healthy volunteers. Chemotherapy 42(6):402–409
13. Shepard RM, Duthu GS, Ferraina RA, Mullins MA (1991) High-performance liquid chromatographic assay with electrochemical

detection for azithromycin in serum and tissues. J Chromatogr 565(1–2):321–337

14. Sphon JA (1978) Use of mass spectrometry for confirmation of animal drug residues. J Assoc Off Anal Chem 61:1247–1252

15. Wallhäuser KH (1967) Antibiotika. In: Stahl E (Hrsg). Dünnschicht-Chromatographie. Springer Verlag. Berlin. S. 545

INDEX

S

T

V

W

X